Accelerated Materials Discovery

Also of Interest

Data Science in Chemistry.
Artificial Intelligence, Big Data, Chemometrics and Quantum
Computing with Jupyter
Thorsten Gressling, 2021
ISBN 978-3-11-062939-2, e-ISBN 978-3-11-062945-3

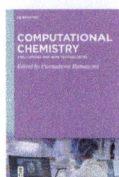

Computational Chemistry Methods.
Applications
Ponnadurai Ramasami (Ed.), 2020
ISBN 978-3-11-068200-7, e-ISBN 978-3-11-068204-5

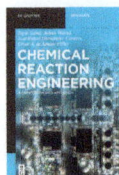

Chemical Reaction Engineering.
A Computer-Aided Approach
Tapio Salmi, Johan Wärnå, José Rafael Hernández Carucci
and César A. de Araújo Filho, 2020
ISBN 978-3-11-061145-8, e-ISBN 978-3-11-061160-1

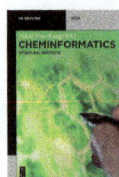

Chemoinformatics of Natural Products.
Volume 1: Fundamental Concepts
Fidele Ntie-Kang (Ed.), 2020
ISBN 978-3-11-057933-8, e-ISBN 978-3-11-057935-2

Maths in Chemistry.
Numerical Methods for Physical and Analytical Chemistry
Prerna Bansal, 2020
ISBN 978-3-11-069531-1, e-ISBN 978-3-11-069532-8

Accelerated Materials Discovery

How to Use Artificial Intelligence to Speed Up
Development

Edited by
Phil De Luna

DE GRUYTER

Editor
Dr. Phil De Luna
2709-1001 Bay St.
Toronto ON M5S 3A6
Canada
p.deluna91@gmail.com

ISBN 978-3-11-073804-9
e-ISBN (PDF) 978-3-11-073808-7
e-ISBN (EPUB) 978-3-11-073325-9

Library of Congress Control Number: 2021952147

Bibliographic information published by the Deutsche Nationalbibliothek
The Deutsche Nationalbibliothek lists this publication in the Deutsche Nationalbibliografie;
detailed bibliographic data are available on the Internet at http://dnb.dnb.de.

Preface

Passion for discovery is at the core of any successful scientist. The nature of research is to push boundaries and increase the sphere of human knowledge, even if by an inch. Throughout their career, every researcher has their own eureka moment – when after hours of doing the same laboratory experiment, or rewriting the same code, over and over again, then all of a sudden – bang! Somehow, someway, it works. That journey of empirical trial and error has marked how science has been conducted since the dawn of time. But today that era of discovery is coming to an end.

There have been three critical advances in technology over the past few decades that have revolutionized the way we discover new things. First, computational power has exploded. Today we carry more computing power in the cell phones in our pockets than in the rockets that brought man to the moon. This computing power has allowed humans to transcend biological limits of brain matter to tackle ever increasingly complex and wicked problems. Second, data is everywhere. We live in a data-rich world that is only getting richer. Advances in the way we collect, analyze, and store data has led to an exponential growth of information. Data is the new gold, and there's a rush by technology companies looking to capitalize and extract full value from this untapped resource. Third, robots are better, much better. Right now, you can go online and purchase a robot dog from Boston Dynamics, one that can perform routine maintenance on oil rigs or explore mining caverns that are unsafe. Robotics has never been cheaper and more accessible as today, and as economies of scale mature, this is only going to drive adoption and innovation.

Taken together, these enabling technologies are accelerating the way scientists discover new things – from drugs and chemicals to batteries and consumer electronics. Automation and artificial intelligence (AI) are being applied to augment and improve traditional discovery. These efficiency gains are critical to saving humanity's most precious resource – time. In a world challenged by a global pandemic and a warming climate, we cannot afford the one or two decades for the traditional cycle of discovery to impact. We need to act now.

This book is a primer for any materials scientist looking to future-proof their careers and get ahead of the disruption that artificial intelligence and automation is just starting to unleash. It is meant to be an overview of how we can use these digital technologies to augment and supercharge our abilities to discover new materials. Our enticing offer is twofold – speed and reproducibility.

The first chapter will provide an introductory overview to accelerated materials discovery. It discusses the basic concepts and levels of automation, the role of machine learning and artificial intelligence, the importance of data, the experimental workflow, and what the laboratory of the future will look like.

The second chapter describes how artificial intelligence is being used to model materials chemistry with catalysis for clean energy conversion as the application goal. It provides basic concepts of chemical reactivity and how we can describe these

https://doi.org/10.1515/9783110738087-202

properties digitally and in a machine-readable way. It then shows how applying machine learning to computational models of different catalysts can help us discover new ways to convert CO_2 into fuels, make hydrogen, or transform methanol.

The third chapter tackles how we can use artificial intelligence to help us better experimentally measure, characterize, and probe the properties of materials. Spectroscopy investigates materials through electromagnetic stimulus, x-rays or light. The amount of data generated by spectroscopy measurements are massive, and the analysis is complex. By using artificial intelligence, we can speed up the analysis and find complex correlations in the sea of spectroscopy data. This chapter tells us how.

The fourth chapter covers the concept of self-driving laboratories. Autonomous labs that plan, conduct, and draw conclusions from experiments with minimal human input. Robotic labs have been science fiction for many years, but now the advances in robotics, AI, and data are allowing these autonomous researchers to come to life.

The fifth chapter dives deep into what makes artificial intelligence intelligent – the complex math and algorithms that give AI its power. This chapter will cover the various types of algorithms used in materials discovery including neural networks, natural language processing, and unsupervised learning. It describes the strengths and pitfalls of each and what remains to be done to improve them.

The sixth chapter is written from the perspective of how accelerated materials discovery can be used effectively in industry. Authored by a leading scientist in one of the world's foremost materials informatics startups, this chapter ties all the concepts together and provides case studies for how and why companies should use accelerated materials discovery to remain competitive.

The intent of this book is to provide a foundation for others to build, and it is my sincerest hope that this book becomes obsolete very quickly. I hope you find it useful and engaging as you continue your journey to expand the sphere of human knowledge, if only by an inch.

October 2021
Phil De Luna

Contents

List of contributors

Dr. Robert Black
National Research Council of Canada
Energy, Mining, and Environment
Research Centre
Mississauga, ON, L5K 1B1
Canada
Robert.Black@nrc-cnrc.gc.ca

Dr. Isaac Tamblyn
National Research Council of Canada
Security and Disruptive Technologies
Research Centre
Ottawa, On, Canada, K1N 5A2
Canada
Isaac.Tamblyn@nrc-cnrc.gc.ca

Prof. Samira Siahrostami
University of Calgary, Department of
Chemistry
2500 University Drive NW
Calgary Alberta T2N 1N4
Canada
samira.siahrostami@ucalgary.ca

Dr. Stanislav Stoyanov
University of Alberta
Chemical and Materials Engineering Dept
116 St. and 85 Ave., Edmonton, AB
Canada
stoyanov@ualberta.ca

Dr. Sergey Gusarov
National Research Council of Canada
Nanotechnology Research Centre
11421 Saskatchewan Dr NW
Edmonton, AB T6G 2M9
Canada
Sergey.Gusarov@nrc-cnrc.gc.ca

Prof. Ian D. Gates
University of Calgary
Department of Chemistry
2500 University Drive NW
Calgary Alberta T2N 1N4
Canada
ian.gates@ucalgary.ca

Mohammadreza Karamd
University of Calgary
Department of Chemistry
2500 University Drive NW
Calgary Alberta T2N 1N4
Canada
mohammadreza.karamad@ucalgary.ca

Steven B. Torrisi
Harvard University
Department of Physics
17 Oxford Street
Cambridge, MA 02138
United States of America
torrisi@g.harvard.edu

Prof. John Gregoire
California Institute of Technology
1200 East California Boulevard
Pasadena, California 91125
United States of America
gregoire@caltech.edu

Dr. Junko Yano
Molecular Biophysics and Integrated
Bioimaging Division
Lawrence Berkeley National Laboratory
1 Cyclotron Road
Berkeley, CA 94720
United States of America
JYano@lbl.gov

Matthew R. Carbone
Department of Chemistry
Columbia University
3000 Broadway
New York, NY, 100027
United States of America
mrc2215@columbia.edu

Prof. Carla Gomes
Cornell University
Department of Computer Science
107 Hoy Rd
Ithaca, NY 14853
United States of America
gomes@cs.cornell.edu

https://doi.org/10.1515/9783110738087-204

Dr. Linda Huang
Toyota Research Institute
4440 El Camino Real
Los Altos, CA 94022
United States of America
linda.hung@tri.global

Dr. Santosh Suram
Toyota Research Institute
4440 El Camino Real
Los Altos, CA 94022
United States of America
santosh.suram@tri.global

Benjamin P. MacLeod
University of British Columbia
Department of Chemistry
2036 Main Mall
Vancouver, BC, V6T 1Z1
Canada
ben.macleod@berlinguettegroup.com

Fraser G. L. Parlane
University of British Columbia
Department of Chemistry
2036 Main Mall
Vancouver, BC, V6T 1Z1
Canada
fraser@berlinguettegroup.com

Amanda K. Brown
University of British Columbia
Department of Chemistry
2036 Main Mall
Vancouver, BC, V6T 1Z1
Canada
amanda@berlinguettegroup.com

Prof. Jason E. Hein
University of British Columbia
Department of Chemistry
2036 Main Mall
Vancouver, BC, V6T 1Z1
Canada
jhein@chem.ubc.ca

Prof. Curtis P. Berlinguette
University of British Columbia
Department of Chemistry
2036 Main Mall
Vancouver, BC, V6T 1Z1
Canada
cpb@berlinguettegroup.com

John Dagdelen
Univeristy of California Berkeley
Berkeley, CA 94720
and
Lawrence Berkeley National Laboratory
1 Cyclotron Rd. Berkeley, CA 94720
United States of America
jdagdelen@berkeley.edu

Alex Dunn
Univeristy of California, Berkeley
Berkeley, CA 94720
and
Lawrence Berkeley National Laboratory
1 Cyclotron Rd. Berkeley, CA 94720
United States of America
ardunn@lbl.gov

Dr. Gustavo Guzman
Citrine Informatics
2629 Broadway
Redwood City
CA 94063
United States of America
gguzman@citrine.io

Robert Black, Isaac Tamblyn

1 An overview of accelerated materials discovery

1.1 Accelerated materials discovery and the promise of autonomous material discovery

New materials are core to forward progress in technology. We refer to historical periods based on the prevalent working material of the time: the Stone age, the Iron Age, the Bronze Age, and so on. Major technological leaps are often closely related to new approaches to material synthesis, processing, or creation. Record setting vehicles often operate at the extreme of material limits (e.g. titanium in the world's fastest manned aircraft (SR-71), carbon fiber and Kevlar in high performance racing sails, pristine and ultra-pure materials for semiconductors for solar cells in spacecraft). As the rate of technological improvement increases, so does the demand for research to find new and better materials and material processing capabilities. Human-centric research is becoming increasingly expensive to perform, and scientific breakthroughs are becoming less efficient in terms of researcher productivity and costs compared to past innovations [1]. Factors pushing research in more expensive directions (both monetary and time requirements) are numerous: the magnitude of data being generated is increasing substantially year after year, expenses associated with sophisticated and specialized equipment to understand more complex and ground-breaking ideas becomes increasingly more expensive, and the resources required to know and understand all the information available is reaching a critical point. This increased complexity has resulted in timeframes of 20+ years for a product to go from conception to market [2]. Naturally, with the need for further investments to push R&D efforts forward, it is of no surprise that investors shy away from "deep technology" for investment return, that which focuses on material and/or hardware production, and instead is moving towards a more digital space [3].

Robert Black, Energy, Mining, and Environment Research Centre, National Research Council of Canada, Canada
Isaac Tamblyn, Security and Disruptive Technologies Research Centre, National Research Council of Canada

https://doi.org/10.1515/9783110738087-001

1.1.1 Accelerating experimental workflow

To accelerate research efforts, traditional human-centric experimentation has a significant opportunity to undergo a paradigm shift. In the world of materials, this can mean designing, developing, and deploying automated experimental platforms (laboratories) that perform synthesis and characterization, with the goal of using artificial intelligence and machine learning (AI/ML) to optimize a specific material property, either through a more efficient "exploratory" method, or through capabilities to explore new design space that was previously not considered. This shift will come through the application and integration of one or multiple of the following areas into the experimental workflow:

1. Automated experimentation – Robotic manipulation of benchtop experimentation, data handling and operation of equipment, integration of various experimental pieces into a unified platform capable of serial or high-throughput operation
2. Artificial intelligence/machine learning – Utilization of data to train statistics-based models to determine the optimal experimental space towards a specific material function, including concepts of inverse material design
3. Data – A unified protocol for the collection and storage of data. This includes data acquisition from the measurements, post-processing and curation, storage of relevant information into various databases, and assurance of data accuracy and integrity
4. Computer simulation – Augmenting the experimentation and AI with first principles or simulation-driven aspects of the material of interest

The combination of these concepts enables accelerated material discovery, with the exact degree of integration of each concept dependent on the desired goals and the level of autonomy required of the experimental platform. We take an analogy to that of the Society of Automotive Engineers (SAE) International definition of autonomy for self-driving cars [4] where different levels of autonomy are defined based on the integration and advancement of the components that comprise an autonomous researcher (Fig. 1.1).

–   **Level 0 – base level.** A typical human-centric experimental process and/or current-day laboratory. The majority of experimentation is carried out by humans. Some pieces of equipment are used and provide signals in digital formats, but there is essentially no exchange of information other than what is measured and processed by a human. As an example, a device might collect data and provide some peak positions, but the output format is usually an image which ends up in a scientific publication, with no outside data transfer.
–   **Level 1 – human assistance.** This sees the implementation of some type of assistance to the human tasks, whether physical or digital. Assistance is minimal and confined to only a single task, such as the use of pipetting machine and/or some form of automated data collection. At this stage, one can expect the

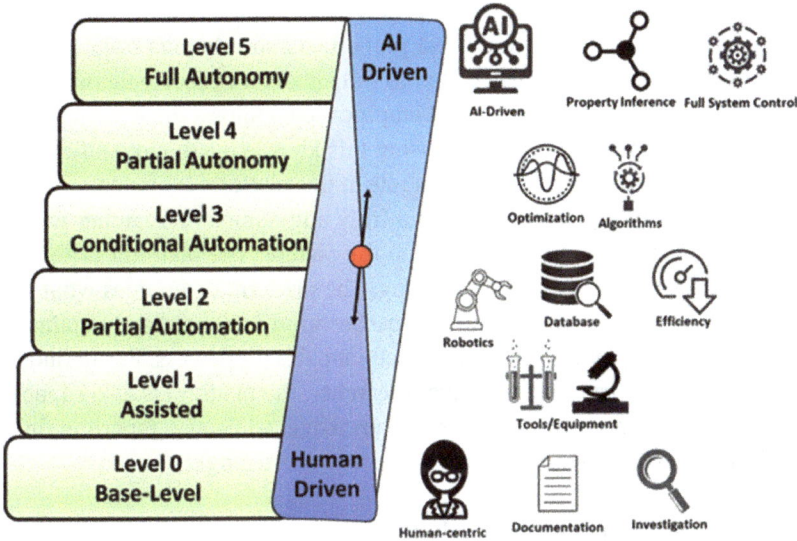

Fig. 1.1: Five levels of an accelerated experimental workflow, with each successive level integrating more components to eventually reach full autonomy. The red dot marks the threshold for crossover from experimental automation to experimental autonomy.

acceleration factor of experimentation to increase and the experimentation protocol to be more efficient, but still the workflow is entirely dictated by the decisions of the human.

– **Level 2 – partial automation.** At this stage, the experimental workflow is becoming much more automated, with various experimental procedures capable of operating over extended periods of time. This could be considered a "high-throughput" stage, as the workflow pace greatly increases. While the workflow is enhanced through the incorporation of either physical or digital means, the human is still in control of the experiments being run, with their own analysis and decisions controlling the workflow of the platform.

– **Level 3 – conditional automation.** This stage sees the application of AI/ML to create optimization loops which are coupled to the platform automation. Through the incorporation of statistical and/or deep learning algorithms into the workflow, the system will be capable of making its own decisions based on the empirical data that is produced during a single campaign. This can be considered the base level for a self-driving laboratory. While direct human involvement in the process is reduced compared to previous levels, domain knowledge is still key to successfully drive the experiments effectively. This includes the human defining the experimental features, and scientific intuition to develop the proper workflow is still necessary.

- **Level 4 – partial autonomy.** At this stage, the system starts to truly become autonomous, as it begins to infer chemical properties and relations from outside sources (literature, past experiments, etc.) to truly plan and execute autonomous experimentation. Coupling of theory and empirical data allows the system to understand and infer chemical processes, using this knowledge to more efficiently search the material landscape. The human role in this platform is limited.
- **Level 5 – full autonomy.** This indicates a truly autonomous researcher where the human can effectively be removed from the picture. The platform is able to complete all tasks of discovery and reporting. The platform determines what experiments, simulation, and theoretical approaches need to be applied to optimize the active learning algorithm in a truly closed-loop design process. Furthermore, this stage represents a truly autonomous researcher, the platform is also capable of reading the literature to update its own knowledge, while also reporting findings in a human readable format (natural language processing).

As technology advances and we can scale the "autonomous material discovery" ladder, we become closer to realizing the concept of a truly autonomous researcher. This is the required paradigm shift to greatly accelerate the pace at which deep-technology experimentation is performed, while promising significant cost reduction and time savings for the material development process [5]. This paradigm shift has been recognized by others, in particular the role that big data and data-driven processes will move research into a new paradigm [6].

In what follows, we will discuss the role laboratory automation, machine learning, and data infrastructure will play in reaching truly autonomous experimentation, our view of where the field is moving in the near and the medium term, and where the opportunities are for new advancements in this area. The field of accelerated material discovery is currently only at its infancy and is an exciting advance in the history of scientific discovery – the beginning of a new era for material discovery, design, and control.

1.2 Laboratory automation to enable machine learning and artificial intelligence

Our discussion will begin with laboratory automation, as this is the key enabler of data acquisition and is the heart of any self-driving laboratory for accelerated material discovery. Laboratory automation refers to the coupling of physical experimentation via robotic control with software to perform repeated tasks. Much akin to a world-class researcher working in the lab, automated experimentation is key in ensuring a consistent, non-variant workflow that is theoretically able to run for days on end, collecting useful and meaningful data throughout the entire experimental campaign. Automated experimentation, has changed and become more widely adopted over the years as new concepts and tools are developed to advance research capabilities beyond traditional human-centric approaches. Regardless of all advances, the purpose has remained the same – decrease the resources and time required for experimentation, while ensuring a consistent protocol which adheres to strict data collection and curation standards.

1.2.1 Combinatorial experimentation

To aid in the discussion of automated experimentation for accelerated material discovery, it would be remiss to not touch on the concept of combinatorial experimentation, and how this methodology greatly improves the throughput of experimentation compared to traditional human-centric methods. Some of the first applied concepts of combinatorial experimentation came to light in the late 1960s and early 1970s as a method of high-throughput experimentation and data collection [7, 8]. Combinatorial experimentation looks to multiply and parallelize multiple components of synthesis and analysis in a single step, thereby vastly reducing the time for experimentation by factors of up to 1000 due to parallel design and characterization. The hope is that the optimal material lays within the explored space, and it is just a question of landing in the correct spot to discover the best material. This method has since been applied to various aspects of material discovery, finding its home in the areas of drug discovery [9–11], thin-film production [12–14], and energy-related materials [15, 16].

Initially, combinatorial chemistry was met with great praise and promise, but a critical look at what was achieved with this methodology brings into question the efficacy of such a process and whether the promise of this approach as an efficient material discovery process was actually realized [17, 18]. In particular, the concept of creating vast libraries of chemical compounds and compositions became quite infeasible when scaled up beyond that of small molecules, for even with the accelerated synthesis and characterization methods, the available design space for new material discovery becomes exponentially large as more permutations of material "combinations" are introduced [19]. Thus, combinatorial chemistry is ideal when exploring small molecule/material space and for the generation of highly focused libraries as a means for material optimization. Using this method from the beginning of a material exploration process, in particular for large potential material spaces, will result in wasted resource space without any actual optimization having occurred. Combinatorial experimentation is ideal if the goal is to acquire a significant amount of data covering a vast experimental landscape in a rapid timeframe (high-throughput). Automation of this process further reduces the workflow timeline and typically will save on material resources. However, combinatorial chemistry has its flaws – in particular when it comes to the concept of self-driving laboratories. These include, but are not limited to:

- Specialized equipment is needed, often with significant up-front costs for initialization
- Large heterogeneous datasets are produced
- Wasted optimization space (i.e. random controlled search)
- Resource intensive and complicated use of AI/ML for such highly dimensional experimental protocol
- Potentially massive exploration space – infeasible to achieve within a lifetime

Regardless, this field of combinatorial experimentation is still advancing as new technologies are produced. The application of computation and theory is making great strides in the area of combinatorial chemistry as a promising avenue forward, in particular to down-select material libraries and thus vastly shrink the exploration space for experimentation to address the issue of wasted experimental resources.

1.2.2 Combinatorial experimentation vs. accelerated materials discovery

When we compare the history of combinatorial chemistry to the hype surrounding accelerated material discovery today, what makes the current push for accelerated material discovery different? Are we simply making the same claims of 30 years ago, only to be unfulfilled in what this promise of accelerated material discovery produced in the future? There are some key differentiators that have taken place in the past 30 years that provide hope in this area of self-driving laboratories and data-driven accelerated material discovery doesn't meet the same fate:

1. The widespread availability of high quality, well documented, and easy-to-use open-source software. This greatly opens the barriers for who can access and utilize this concept in their own research.
2. The significant and steady increase in the computing power which is available at the benchtop and departmental level, creating ever-increasing capabilities to process and learn from more data.
3. The increased knowledge of programming and data-science among students and reseaerchers.
4. The rapid increase and profile of AI tools.
5. The significant reduction in cost of microelectronics, sensors, and automated laboratory equipment, enabling research from all levels able to apply automated experimentation to accelerate their workflow.
6. The maker movement culture. More than ever there is a desire to apply, create, and discover new processes and tools to make experimentation more efficient.

When it comes to self-driving laboratories, the goal is to move from the high-throughput experimentation approach that combinatorial chemistry offers into an area of intelligent-throughput experimentation, which will be the focus for the remainder of this chapter. Intelligent-throughput experimentation is the use of laboratory automation to enable machine-learning algorithms (in particular, active learning) to select the next experiment (or batch of experiments) based on the collected experimental data in real time. Such an approach is more amenable to serial synthesis and characterization protocols as opposed to excessive parallelization, minimizing the actual number of experiments that are performed to find an optimal material property. Perhaps more importantly, with this type of experimentation enabling the use of AI, this approach is much more appropriate to use for "from the beginning" types of material exploration without any bias or prior knowledge of the material space – successfully finding the optimal material properties without the need for the mass experimentation required by combinatorial chemistry. Numerous publications have benchmarked the results of this autonomous process versus experts and other design of experiments methods, and in all cases machine-learning optimization demonstrates a more efficient material discovery process [20–22]. Figure 1.2 is a representation of how grid search, algorithm-based (active learning), and human-centric methods of exploration have been shown to compare based on various metrics per number of experiments.

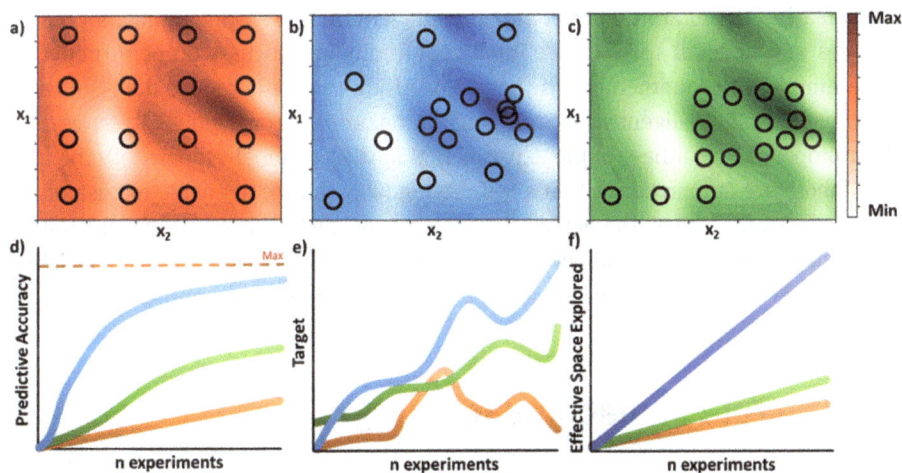

Fig. 1.2: A hypothetical optimization space of variables x1 and x2, with the representation of samples by black dots utilizing a) grid search, b) active-learning algorithm and c) human-based search methods. Metrics of performance vs. number of experiments are shown in d) predictive accuracy towards a global maximum, e) the actual performance target of each experiment, and f) effective space explored (how effective is the search towards reaching a global maximum). This figure was inspired by results from [20, 22].

Similar to a single researcher working manually in the lab, to date, the use of self-driving laboratories for accelerated material discovery has focused on running a more serial (or small subset of parallel) experimental process, but with great focus on quality of data – consistency, reproducibility, and interpretability. This information is then utilized to intelligently select the next experiment. This contrasts the previously discussed high-throughput or combinatorial approach. In comparison, the benefits of this intelligent throughput type of laboratory automation include:

- Reduced investment costs – can even be performed with normal benchtop equipment assuming the technical resources are available [23]
- More streamlined focused workflows that better mimic true laboratory settings (i.e. the workflow can be more flexible as is the case of a researcher working in the laboratory)
- Smaller more manageable number of "focused" datasets, thus helping to ensure homogeneity in the dataset
- Modular process that can more easily be outfitted with new/change in equipment to develop or improve workflows
- Reproducibility across laboratories – allowing for meaningful sharing without expensive duplication of measurements

Note that the key to selecting the right approach (high-throughput or intelligent throughput) is dependent on the end goals. High-throughput experimentation focuses on increasing the total number of experiments within a given time-frame, sweeping across a vast material space, while intelligent-throughput focuses on performing the same number of experiments with the use of AI/ML to greatly reduce the number of experiments and intelligently navigate towards maximizing some global objective (material property). Regardless of the experimental approach (serial vs. parallel), the goal with laboratory automation within a self-driving laboratory is to enable the use of AI/ML to effectively "drive" the exploration process. As such, it is critical for laboratory automation to create consistency in data collection and possess a highly consistent and controlled workflow as to enable the use of machine-learning approaches.

1.3 The role of machine learning

With the previous discussion on experimental automation, outside of improving the experimental throughput, one of the main reasons to utilize automation is to enable machine learning. Machine learning has many potential uses in self-driving laboratories, and an entire discussion on how machine learning can be applied to accelerated material discovery is extremely vast and outside the scope of this chapter. This section will focus on a subset of important concepts/principles of machine learning, in particular those related to experimentally driven material discovery. A discussion of how machine learning is applied to self-driving laboratories, its critical importance to accelerate the material discovery process, and how machine learning is influenced by various other aspects of the platform to create an autonomous researcher.

At its heart, machine learning is essentially the development and use of algorithms to find and apply patterns within data to develop a model which best represents that data, and thus can be used to predict or infer the behavior of new, previously "unseen" data. This data-driven approach is different compared to other model methods in the physical sciences, which typically rely on an underlying theoretical framework coupled with computation and simulation (DFT, quantum chemistry calculations, etc.). Machine learning relies heavily on statistical learning and stochastic processes. It is a data-driven process typically grounded in the areas of variability and uncertainty that identifies underlying patterns and correlations within real data without the need for underlying physical science or theoretical framework. In today's world, some of the most prominent uses of machine learning are in the areas of image and speech recognition, recommendation algorithms, medical diagnoses, and forecasting within the finance sector.

Machine learning has typically been defined to come in three different categories: *supervised, unsupervised, and reinforcement learning.*

1.3.1 Supervised learning

Supervised learning refers to the use of labeled data to train the ML model. In the material discovery process, this would be using a dataset where each data point has a known label or some form of identification. These labels can be categorical in the case of classification problems or scalar/continuous values for regression problems. The machine learning algorithm seeks to find a function (the form of which may or may not be known, but is learned from the training data) which can accurately map x to values of y for unseen or newly introduced (x,y) pairs. Traditional curve fitting which is typically taught in an introductory science course is a supervised learning problem. Generative models are a subclass of supervised learning, but they are typically described separately because the learning task is in some sense "opposite" to classification. As an example, rather than identify the presence of a cat within an image, a generative model must generate new images or sounds of cats which are different from those within the training set, yet close enough that they still belong to the same class. Mathematically, the objective of a generative model is to produce a sample of $p(x)$. As this relates to materials, this would be to identify or propose new materials based on a previously learned subset of material properties and attributes. Supervised learning is overwhelmingly the most popular use of machine learning in the area of new material discovery, as it forms the basis of using collected experimental (or theoretically produced) datasets to generate models that are able to predict new material properties, aid in the characterization process of materials, or optimize experimental approaches (design of experiments).

1.3.2 Unsupervised learning

Conversely, unsupervised learning is concerned with datasets where there are no labels. The dataset has a variety of material features and measurements but no indication of the identity or type of material. Typically, the goal of unsupervised learning is to find similarities and differences within the data to be able to label existing and new data as a means to elucidate understanding. Unsupervised learning is closer to the learning process of humans and other intelligent species – most of the observations we make throughout our lives are not accompanied with an external labelling signal. One of the key differentiators of intelligent behavior is the ability for higher organisms to detect patterns in unstructured datasets, a direct application of unsupervised learning.

1.3.3 Reinforcement learning

Reinforcement learning (RL) is a separate concept all-together based on trial and error, where the model learns and adapts based on its own actions through a learning agent. RL agents are capable of learning to control dynamical systems through experience. Initially, an RL agent will have no data and no understanding of the approach/ problem. Through interaction with an environment through a series of sequential actions, a RL agent will receive observations of the environment's state and associated rewards based on the selected action (did it perform well or poorly based on the user's definition of success). RL agents seek to maximize the discounted sum of rewards. Real-world examples of reinforcement learning exist in the areas of robotics and gaming AI [24, 25].

1.3.4 Active learning

Note that the above discussion of machine learning definitions is not rigid, and this constantly evolving field rapidly brings about new developments. In particular, a type of semi-supervised learning approach known as active learning is a critical concept when it comes to self-driving laboratories. Supervised learning models are trained through taking a random sampling of representative data, and using this data to generate a model which then best fits the data through the maximization of a particular objective function. One could say this approach is *passive learning* as it is reliant on the large amount of randomly sampled data. Active learning, however, gives the algorithm agency to *choose* the data points to sample and generate the model. The algorithm starts with a surrogate machine-learning model and a small sampling of data, and based on the responses from these data points, uses this to predict (query) the next sampled data point that will provide the greatest knowledge to update the current model. The criteria used to query the next data point is dependent on the model being produced, but typically is in the form of an acquisition function looking to reduce the least confidence, or spread, in the variability of a specific region of the objective function. The surrogate model will be continually updated to fit the true optimization landscape, and thus finds the solution to maximize (or minimize) the objective of the design space. This approach has been shown to be highly successful when implemented with a robotic platform to create a self-driving laboratory, able to successfully guide the experimentation to find material compositions that exhibit an optimal property [21, 22, 26–31]. This approach is loosely reminiscent of typical human-approach laboratory experimentation, and thus why the concept of active learning is critically important to the accelerated material discovery process, in particular when it comes to the development of models involving empirical data where the size of the dataset is sparse (compared to theoretical or computational datasets). While not offering an abundance of data compared to the high-throughput methods, the active learning approach will be more efficient at updating the model in terms of time, cost, and resources, and thus is an extremely efficient (and powerful) machine learning method when applied to experimental material discovery processes.

1.3.5 Machine learning vs. deep learning

Another important aspect of machine learning which encompasses the various classifications above is the concept of feature-based machine learning versus deep learning, both of which can play a role in accelerated material discovery. Feature-based machine learning finds patterns and correlations based on user-defined features (also called descriptors) that make up the dataset. From a material perspective, this is making a model to describe the dataset based on various material attributes and characteristics defined by the user (varying from just a few features to large dimensionality if the information is available). As these features are selected by the user and influenced with domain knowledge, these features can be rooted in scientific understanding and interpreted based on physical or chemical properties along with empirical data from some type of experimental response. The features are preserved within the model, and based on the model can potentially be used to infer physical and scientific processes of the material of interest [32–36]. Featured-based machine learning algorithms are rooted in statistical methods such as linear regression, k-nearest neighbor, decision trees, and Gaussian processes. Neural networks and deep learning, on the other hand, are focused on developing models to extract and identify features from raw data. That is, the features are extracted as part of the training process in order to generate the best model to represent the data. Such an approach is extremely powerful for developing models, but the "black-box" nature of modeling non-linear relationships in data makes it difficult to interpret the model for scientific understanding and reasoning, as the features generated to train the model can be quite abstract and not rooted in an scientific understanding.

The selection to use feature-based machine learning or deep learning in material discovery is dependent on a variety of factors. As a generalization, feature-based machine learning is computationally inexpensive, and more interpretable for relating patterns/relations to scientific understanding and/or mechanistic processes [37]. Deep learning, on the other hand, does not require feature engineering or even *a priori* knowledge about the dataset. Objectively, neural-network approaches can find patterns/relationships in data, in particular non-linear relationships (even though it may be more "black box" in how it does so), and while it typically requires more available data to generate useful models than feature-based machine learning, it can also be more readily applied to various data forms given its featureless nature (images, spectra, etc.) [34, 38]. To date, the use of feature-based machine learning methods for material discovery has been used in many areas related to experimental data, including augmented with computational and theoretical data, as well as utilized in self-driving laboratories. Deep learning has mostly been utilized thus far only for computational and/or theoretical material discovery or where generally the generation of data and the type of analysis allows for deep learning.

1.3.6 Applying machine learning to automated labs

The application of machine learning to a self-driving laboratory can vary depending on the needs of the researcher and the overall goal of the platform. However, the role of machine learning is to make the robots act as an intelligent researcher, determining the needed data to better optimize the objective function and then acting to gather this information (e.g. finding a new material with a specific set of properties) – the active learning approach described previously. To date, common utilization of machine learning falls into the areas of simplifying existing workflows and data analysis (e.g. finding peaks in spectra), design of experiments, and finding patterns and correlations where the use of traditional models is not possible (expensive or complicated) or the experimental observations are non-linear.

Much of the focus on machine learning above is its use to discover new materials using experimentation to determine the material response "features" as the input towards some clear objective related to the material performance. However, for an accelerated material discovery workflow the use of machine learning is also very important for the *characterization* process of a material. As an example, machine learning can be applied to greatly reduce the time required to post-process experimental data. This is critical if high-throughput is a necessary component of the experimental workflow. Quantitative analysis of XANES spectra is a rather laborious task, traditionally relying on comparing experimental data to computed theoretical data. Through the use of neural networks, researchers have developed analytical protocols and machine learning-algorithms to perform quantitative analysis on experimental XANES spectra, greatly reducing the time necessary for analysis [39]. Image analysis is another area where the use of machine learning can greatly improve characterization capabilities. A great example of this is machine learning as applied to electron imaging. Typically, classification related to material shape, morphology, and size can be quite a laborious process if a true statistical analysis is to be performed manually. Researchers have successfully developed and applied genetic algorithm-based deep learning models and cluster-based machine learning models to successfully determine the shape, size, and distribution of formed nanoparticles with extremely high accuracy [40]. Other work has demonstrated how the application of fully convolutional neural networks can be used to train a model on simulated scanning transmission electron microscopy images (STEM) crystalline lattice, and then apply this model to experimental images to not only identify known defect sites (those included in the training set), but also identify and learn new defect structures from the experimental data [41]. Delving deep into the use of deep learning for electron imaging is outside the scope of this chapter, but refer the reader to an excellent review paper on this topic [42]. This approach demonstrates that machine learning/deep learning application to characterization is able to not only assist in the characterization process, but potentially elucidate important information that is missed and/or not initially considered for analysis. Furthermore, the application of these models brings about some automation in the analysis, thus allowing for the implementation of certain characterization methods (like electron imaging analysis) into an automated or high-throughput system – a concept which previously would not be possible and will in the future allow for more sophisticated characterization techniques to potentially be incorporated into self-driving laboratories.

Machine learning can also be used to address the issues of experimental data scarcity, a typical concern when using experimental data compared to theoretical or simulated data. Oviedo *et al.* demonstrates this concept with the generation of a model to identify thin-film crystal structure utilized spectral data from both simulation and experimentation, coupled with a physics-informed data augmentation to expand the training set from 115 XRD patterns to 2000 + patterns [43]. This is used to train a classification model which can serve as a robust framework for future material classification based on an XRD pattern, but also using data coarsening determined that XRD pattern acquisition time could be reduced by 75% with more coarse experimental parameters while still providing precise enough measurement information. The capability of ML to not only assist in the characterization but augment the data to reduce the number of needed experiments (reducing both human resources and cost) is an exceptionally powerful tool.

Finally, as will be discussed further, the credibility and reproducibility of the data are critical to any machine learning model and self-driving laboratory. This concept applies to both the generation of raw data, as well as the methods for material characterization and analysis. Using machine learning for material characterization removes any precognitive bias and/or human influence in the characterization (feature determination) process, and thus ensures the data generated is traceable to a specific application of a model or transform on the original data. While the application of machine learning to data involves stochastic processes, and thus variation in results is expected, this approach will also allow other researchers to reproduce your interpretation of data and not leave understanding up to ambiguous human-interpretable analysis methods. Such normalization of characterization methods, or traceability of processes, will be critical as metadata for future material databases and sharing of data among researchers.

1.4 The importance of data

The previous discussions around automated experimentation and machine learning revolve around the acquisition and use of data. Automated experimentation enables the acquisition of data in a non-biased, potentially higher-throughput fashion, whereas machine learning is entirely based around algorithms which utilize said data to drive the discovery process. Thus, to understand the accelerated material discovery ecosystem, the importance of data and various concepts surrounding the acquisition and storage of data must be introduced. Here, we will briefly touch on the importance of data as it relates to accelerated material discovery, mostly through an academic/small organizational lens as it applies to the material discovery process and provide insight into the current state of scientific data and the direction data collection, curation, and open-source sharing is headed in the future.

1.4.1 Open science and open data

Traditionally, in the field of materials science data, is only published and made available to the public in the form of research publications or focused research studies. The data are typically presented in a fashion that best supports the results and is directly relevant to the topic at hand. Sharing of data has traditionally been viewed negatively, in particular within industry, as it is assumed that all data need to be kept close and is proprietary. This is a practice that is hard to change [44]. As we start to move more towards a paradigm change focused on data-driven approaches for material discovery, the way we think, use, and assess data must fundamentally change to better support this shift. Data must be shared to support the previous discussed data-driven models, where the accumulation of data from various research centers can provide the framework to find correlations and patterns that are hidden in smaller "group" focused research data banks. This is the concept of open science and open data [45]. Open data refers to data accessibility that can be used by anyone – both those that are well versed in the research area, as well as those that are not experts.

The data are easily available for even those non-tech savvy to access and use. From a larger point of view, it is the concept of more than an isolated research group working with data, but instead a dataset being used across groups and research fields, all in different ways beyond that of the intended purposes. As an example, an experimentalist shares online their results to produce a new material via sol–gel synthesis method, which is verified using analytical spectroscopy techniques like X-ray diffraction or Raman spectroscopy. The detailed experimental procedure and characterized spectroscopic datasets are then used to build theoretical computation models about the material structure by a separate group, while the experimental successes and failures are used to help build machine learning models to optimize the synthesis procedure and/ or discover new materials by another group. This is the concept of open science, which ultimately can lead to cascading scientific explorations compared to the more traditional methods of data dissemination.

A historical perspective of the movement towards open science and data policy is covered in great detail by Himanen *et al.* [36]. To borrow from this perspective, it has become clear that the world is advancing and realizing the need for an open data platform, as evident by the tremendous growth in industry and academics as outlined in the referenced perspective. This has led not only to more data being available, but also represents a change in the *way* data is accessed. This is evident from the terminology used to describe areas of data handling and the advancement of data infrastructure:

Databases – where data are held in a repository for archiving, with search functionality to encourage users to share data

Data center – the incorporation of data analysis tools into the database, encouraging users to not only access the available data but perform their own analysis and interpretation of the data

Discovery platforms – With a focus on data mining and AI/ML development, expand data centers to develop workflows so that researchers can query and mine said data to incorporate into their own workflows. This concept of discovery platforms is of most importance to researchers for accelerated material discovery, as it offers open access to various forms of data, allowing for new discoveries and uses to be made from the data beyond its initial purpose

1.4.2 Benefits of open data

To this end, with the world moving more towards an open data policy, the benefits of having open data access are numerous [46, 47]:

- Accelerates the pace of research through allowing others to build promptly on results
- Improves replicability of data, also acting as a method of scientific integrity and truthfulness in results

- "Failed" data is good data, in particular as it relates to experimental results. Failed experiments typically may not make final publication, but for generating AI/ML models failed experimental data is highly valuable [48]
- Can be used to find incorrect results, or erroneous results due to the method or process used
- It enables scientists to test whether claims in a paper truly reflect the whole dataset
- Improves the attribution of credit to the data's originators
- Enhances or creates new opportunities for learning activities
- Establishes good will between researchers, and increased confidence for collaborative opportunities and data information sharing practices

1.4.3 FAIR data management

With the ultimate goal to achieve autonomous systems, data from external sources must be used to influence machine learning algorithms to make better-informed decisions. As discussed previously, the key to reaching these levels of autonomous laboratories is good data management – in particular, ensuring data from outside sources can be used to inform AI/ML models, as well as trusting the validity of the data that is being used. Keep in mind, that a majority of the data collected to date does not have machine learning-use in mind, hence the data may not be usable for one reason or another. It is clear that for truly autonomous experimentation to be realized, the scientific community must practice good data management. For this reason, many industry and academic leaders have devised a set of guidelines to produce a digital ecosystem for acquiring and storing data, known as the **f**indability, **a**ccessibility, **i**noperability, and **r**euseability principles (FAIR) [49]. These principles are meant to serve as the general framework to curate and archive data so that it can be used and discovered for many years after its initial collection and integrated into a variety of works.

Open and FAIR data is key to ensuring data can make its way into a researcher's workflow; even with this, there is still the question of how to store and curate all of the data. On an individual level, this can start by researchers making their data available from either their own repository or as part of the data-presentation format (included in publication, for example). For researchers, data used in publications should be made readily available in a format that is able to be used by other researchers. This data would be available with associated meta-data, detailing various aspects of the workflow or result production methodology, and not just raw "data-dump" that requires significant preparation prior to use in AI/ML applications. This meta-data varies depending on the type of data. For experimental data, critical information would include synthesis data, classification, preparation/characterization methodology, and the history of the sample (if applicable) needs to be encoded within the provided data.

To this end, one can imagine the ultimate goal for all material science data needs is the creation of a general-purpose repository – a common data schema (material ontology) that can be applied to all experimental data, which can then be stored in an easy-to-access database open for public use. Such a concept of course may never come to pass, as numerous complications and concerns arise with this approach, in particular those around confidentiality and data security. Regardless, there exists numerous examples of publicly available material discovery data centers and platforms with their own approach towards providing an accessible and robust material database, which can serve as excellent resources for how to annotate, curate, and store relevant data [50–52]. What is critical to take away from this discussion is that, as an individual researcher moving into the new paradigm of data-driven material discovery, we urge the need for data history, formatting, integrity, and availability to be part of any scientific workflows moving forward.

1.5 The experimental workflow

The discussion now will shift to the entire workflow, with a view on how the various aspects discussed thus far combine to create an accelerated material discovery process. As was alluded to in our initial discussion on the different levels of autonomous experimentation, ultimately the design of the workflow and the incorporation of various elements will be dictated by the needs of the researcher and the ultimate end-goal of the discovery process. This section will briefly touch on the evolution of the experimental process, as well as serve as a perspective on what technological advances need to occur before truly autonomous levels of material discovery can be reached.

The end-goal of the material discovery process significantly influences how the automation, AI/ML, and data infrastructure is incorporated within the workflow. For example, is the goal of the platform to be one based on acquiring large amounts of data to produce a data repository around a particular compound of materials? Is the goal to achieve a specific material property using the concepts of inverse design and AI/ML to perform experiments in a more intelligent fashion? Starting with the area of robotics and automation, the end-goal of the research will dictate the approach of the experimental automation. For the design of laboratory automation, there is a decision to be made on whether systems are designed for high-throughput experimentation, specifically targeted to perform new tasks and functions towards the end goal, or are robotic arms meant to "mimic" the experimentation of researchers, allowing for a lower-throughput, but more autonomous and modular control, over the experimental process?

1.5.1 Inverse design

To borrow the concepts from an excellent perspective article on accelerating experimental workflow by Stein *et al.* [53], reaching a closed-loop process driven by inverse design involves taking a traditional experimentation workflow and adding both physical and digital/computational "workflow accelerators" to reach successive stages of workflow autonomy. This perspective article provides examples of how these accelerators are applied to the workflow in various real-world examples, addressing issues of scientific and automation complexity and identifying current limitations with scientific complexity and how that will stay within the realm of human experts for the foreseeable future. The concept of inverse design represents a complex material discovery process, but is a new paradigm in the material acceleration workflow, and represents the culmination of various aspects we have discussed previously [54]. As shown in Fig. 1.3, currently, the experimental workflow starts with the chemical space, revealing properties of a particular species or structure (the functional space) via experimentation or simulation. Inverse design reverses this, and instead predicts and identifies the chemical space based on an initial target within the material functional space. This inverse design approach

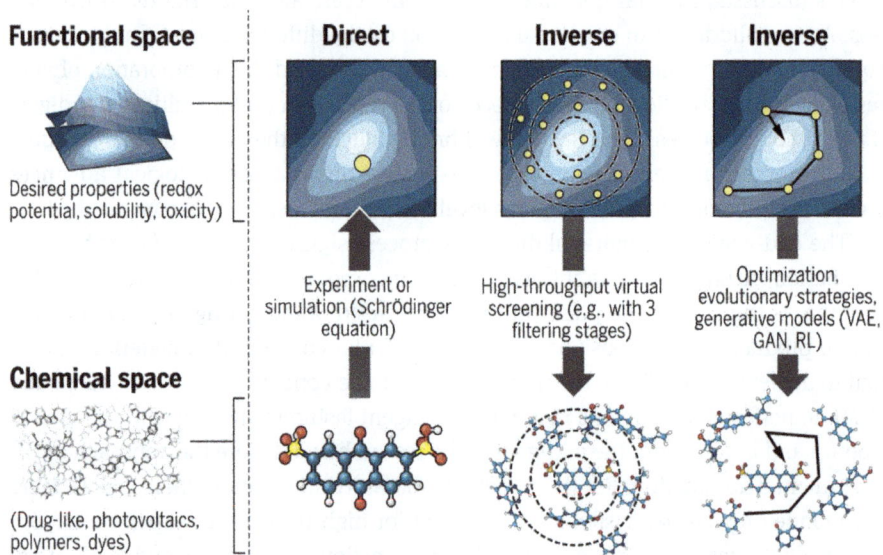

Functional space

Desired properties (redox potential, solubility, toxicity)

Direct

Experiment or simulation (Schrödinger equation)

Inverse

High-throughput virtual screening (e.g., with 3 filtering stages)

Inverse

Optimization, evolutionary strategies, generative models (VAE, GAN, RL)

Chemical space

(Drug-like, photovoltaics, polymers, dyes)

Fig. 1.3: The concept of inverse design as it applies to new material discovery. Traditional experimentation involves the transition from chemical space to functional space for specific material properties. Inverse design starts in the functional space with a desire material functionality, and then explores the chemical space to meet the desired function. Reprinted with permission from [54].

towards material discovery – able to identify a specific property and map this to a specific chemical structure using high-throughput screening and/or generative models – represents an extremely powerful and efficient workflow for the development of new materials targeted towards specific applications. Within the concept of a self-driving laboratory, such an approach is limited by the capabilities to actually synthesize/fabricate materials across the "ideal" chemical space in an automated workflow, with these capabilities only advancing as better technology is developed.

In relation to self-driving laboratories, recent examples use AI/ML to interpret data and plan future experiments based on active learning of "real-time" data collected during an experimental campaign, as discussed earlier in Section 1.3. However, missing from this process is the use of shared information and data acquired outside of an experimental campaign to influence the decision-making, and not rely solely on as-obtained experimental data as inputs. Some recent examples take this one step further and incorporate simulation and/or theory to assist in defining the material objective space [22, 55, 56]. These examples demonstrate how the workflow has developed from purely experimental data to that incorporating generated simulation and theoretical data to assist the discovery process. The next natural step is to utilize historical data directly into the experimental workflow. We have already discussed how existing databases are used in the accelerated material discovery process, but another important, less used component is the data captured in scientific literature. Within the scientific literature is where data forms relationships and its meaning moved beyond raw numbers, and thus it can serve as a critical component for an accelerated discovery process.

1.5.2 Natural language processing

This brings us to a discussion on the use of natural language processing (NLP) and its place as part of the experimental workflow. NLP is a domain of machine learning focused on written languages. Typical NLP problems include machine translation (e.g. French– > English), and predictive text. Like many machine learning disciplines, the rise of neural networks in the early 2010s were transformative for the field. Initial successes based on recurrent neural networks achieved state-of-the art performance in machine translation. Subsequent application of attention-based models (e.g. Transformer) resulted in further improvements. Predictive text has now become commonplace, with many mobile devices and email services providing suggestions for the next word in a sentence, and, in some cases, entire message responses. Beyond these time-saving applications, neural networks applied to NLP have also been used to learn vector representations of words in a structured space which seems to account for their semantic meaning. By training on a large corpus of text, word2vec was able to learn an embedding which allowed for vector operations which revealed a structure of our language [57]. Learned embeddings such as this are useful not only for revealing the structure of our language, but also as a useful starting point for language-based tasks, such as scraping the scientific literature for relevant information to assist in the material discovery process. One example which gained a significant amount of attention was that of Tshitoyan *et al.*, who were able to use the scientific literature to create word embeddings which capture the complex scientific relationships and chemical intuition in the text, and use this information to predict functionality of materials [58]. The authors claim that "latent knowledge regarding future discoveries is to a large extent embedded in past publications," a powerful claim that undoubtedly will be a key component of any accelerated experimental workflow in the future. Weston *et al.* seminal work in this area also demonstrate the capabilities of NLP, using named entity recognition to extract scientific information from millions of journal abstracts [59]. By extracting critical information into an easy-to-query database, this enables a significantly accelerated process compared to manual literature searches and information collection, all of which is made available to the public in an effort of open science as discussed previously [60].

To summarize, at its core, the accelerated workflow follows a typical human-centric approach, involving the generation of a hypothesis, and using experimentation to update and shape the hypothesis and influence the next set of experiments. It is through automation, AI/ML, and data handling to "close the loop" that this process reaches greater levels of autonomy. While shown here as a sequential process, aspects of this workflow can be parallelized for higher throughput, essentially creating various networks and branches of the workflow to further reduce the time of each process. In general, such an approach towards materials discovery lends itself towards experimentation that is typically rapid, allows for complete observation, cost-effective, and ultimately safe to perform in an automated fashion. As an example, solution-based processes involving liquid handling have thus far shown the most success with this self-driving laboratory approach due to the nature of automated solution processes versus a more complicated process for something like solid-state materials, which currently is more limited in automated synthesis tools [53]. However, as this field is only in its infancy, this inverse-design workflow and self-driving laboratory will likely be expanded to further material areas as the laboratory automation technology matures.

1.6 The laboratory of the future

In a world where accelerated material discovery has been embraced and the discovery pipeline starts to aggressively implement various aspects of automation, machine learning, and open data, how does this change the view and function of the traditional research laboratory? What does this mean for laboratories moving forward, and how does the introduction of self-driving laboratories change the current view of the traditional lab? Luckily, we can start to see examples of the "laboratory of the future" through transitions that are currently taking place in industry, with the academic world is now embracing this concept evidenced numerous authors in this book being at the forefront of this movement, among others [61]. Clearly, more traditional lab spaces will have to be "digitized," focused more on automation, communication, and have a robust data handling system to process, curate, and store the increased data flow. This section is meant to only provide a brief overview of how laboratory infrastructure must transform to best utilize self-driving laboratories and serves as a prospective look at what labs will look like 5–10 years into the future as the concept of accelerated materials discovery is gradually adopted.

As the lab becomes more digital, one of the most difficult adjustments will be the change in scientific approach from the scientists themselves. Scientists have been trained for human-centric experimentation, and the change to a more digital landscape that implements accelerated material discovery will require changes in how one thinks about data and the approaches towards experimentation. Just as lab digitization offers new opportunities in the way we perform experimentation, so does it open a new opportunity for multidimensional thinking towards experimental approaches and problem solving. A majority of this is focused around how scientists currently handle and interpret data. As long as data acquisition protocol is followed, there is no such thing as failed data as all data are important for machine learning models. Such an approach to data collection and its use in updating machine learning models may even lead to the answering of questions that the scientist didn't even know existed, a drastic change in approach compared to more traditional research methods [48].

The discussion thus far has focused around the autonomous researcher and digitization of laboratory practices, but what does this mean for the traditional researcher? Theoretically, as technology develops and we move towards a greater degree of research autonomy, the technology will outpace those capabilities of the researcher. Currently, robotics/automation technology is available to vastly out perform a researcher in conducting experimentation. As AI/ML further develops, so too will these algorithms begin to process and correlate more variables, eventually infer chemical processes and properties outside the realm of what is capable with a single human mind. To this end, it brings into question what the role of a human is in these experimental workflows dominated by AI/ML? In all of this, domain knowledge is still king [62]. The workflow design, identification of key experimental descriptors to pursue, and inference plus validity of scientific knowledge from the data linked to physical/chemical principles is still very much the role of the researcher at this stage. Furthermore, within the machine learning itself, integration of domain knowledge has been shown to greatly improve the robustness of models towards unknown data and allow for saved resources when large datasets would normally be necessary [63]. Even if the AI develops to a state where it is able to determine and conduct these various tasks on its own, ultimately the researcher will still play a significant role in the design, build deployment, and execution of the material discovery process. If anything, this process will free researchers from mundane hours in the laboratory and allow more time for the development of ideas and understanding.

Discussed in more detail previously, the experimentation will transform from a human-centric approach to that of a fully automated experimentation platform. This will bring about the development and integration of intelligent sample and data flow (orchestration) allowing for more efficient experimentation enabled by robotics. The degree to which the experimental platforms will interface with humans will depend entirely on the available resources and requirements of the lab. Some experiments, in particular low TRL and "exploratory" work will still require significant human inter- action to allow for maximum flexibility and modularity in approach. Others can be fully executed and driven in an autonomous fashion. On a similar note, the degree to which machine learning is implemented into the workflow will also vary. Cost will play a major role in how labs are designed, as well as the desire for the laboratory to be scalable and flexible; easily adaptable to various research areas and experimental capabilities. The equipment itself will become more universal, and vendors will have to adapt to make their products more flexible for integration into a variety of custom workflow or risk modular laboratories having no desire for rigid equipment.

The data generated from a future laboratory will be in much greater abundance than is generated in traditional laboratories today, and as such the new-age labora- tories will feed into a "data-lake," where all data are curated, stored, and made ac- cessible for necessary handling (whether that be by an individual or AI algorithms). With this, it is key that scientists become data stewards. It is unlikely that scientists will need to be masters in digital data handling/storage and data science, but they will need to understand the role that their data plays in the larger lab ecosystem and how it fits into the data stream. The future of open-source data has already been discussed previously, but to stress that its adoption in academic, government, and potentially even industrial labs is critical to advancing the field of accelerated material discovery. Labs will be much more focused on the development of data- handling ecosystems, with the progress of new approaches and methods for data sharing and garner a more broad-approach towards data analysis. This will shape labs to be much more collaborative in nature, with significant "cross-sharing" of intel- lect and disciplinary expertise. This in turn will lead to new approaches towards experimentation, as the line starts to blur between scientist and engineer and the various workflows become integrated.

References

[1] Bloom N, Jones CI, Webb M, et al. Are ideas getting harder to find?. Natl Bur Econ Res, 2017, 23782.
[2] Eagar TW. Bringing new materials to market. Technol Rev, 1995, 98, 42–49.

[3]   Gaddy BE, Sivaram V, Jones TB, Wayman L. Venture capital and cleantech: The wrong model for energy innovation. Energy Policy, 2017, 102, 385–395.

[4]   Taxonomy and defintions for terms related to driving automation systems for on-road motor vehicles. SAE Int, n.d., J3016, 35.

[5]   SENER-CIFAR-DOE. Materials acceleration platform. Mission Innov, 2018, 108.

[6]   Hey T, Tansley S, Tolle K. The fourth paradigm: Data-intensive scientific discovery. Version 1. Microsoft Corporation, 2009.

[7]   Kennedy K, Stefansky T, Davy G, Zackay VF, Parker ER. Rapid method for determining ternary-alloy phase diagrams. J Appl Phys, 1965, 36, 3808–3810.

[8]   Hanak J. The " Multiple-Sample Concept " in materials research : Synthesis, compositional analysis and testing of entire multicomponent systems. J Mater Sci, 1970, 5, 964–971.

[9]   Sun X, Vilar S, Tatonetti NP. High-throughput methods for combinatorial drug discovery. Sci Transl Med, 2013, 5, 1–8.

[10] Kulesa A, Kehe J, Hurtado JE, Tawde P, Blainey PC. Combinatorial drug discovery in nanoliter droplets. Proc Natl Acad Sci, 2018, 115, 1–6.

[11] Liu R, Li X, Lam KS. Combinatorial chemistry in drug discovery. Curr Opin Chem Biol, 2017, 38, 117–126.

[12] Ludwig A. Discovery of new materials using combinatorial synthesis and high-throughput characterization of thin-film materials libraries combined with computational methods. NPJ Comput Mater, 2019, 5.

[13] Mao SS, Burrows PE. Combinatorial screening of thin film materials: An overview. J Mater, 2015, 1, 85–91.

[14] Joress H, Decost BL, Sarker S, et al. A high-throughput structural and electrochemical study of metallic glass formation in Ni-Ti-Al. ACS Comb Sci, 2020, 22, 330–338.

[15] Yao Y, Huang Z, Li T, et al. High-throughput, combinatorial synthesis of multimetallic nanoclusters. Proc Natl Acad Sci, 2020, 117, 6316–6322.

[16] Park SH, Choi HC, Koh JK, Pak C, Jin S, Woo SI. Combinatorial high-throughput screening for Highly Active Pd – Ir – Ce based ternary catalysts in electrochemical oxygen reduction reaction. ACS Comb Sci, 2013, 15, 572–579.

[17] Carroll J. Will combinatorial chemistry keep its promise?. Biotechnol Healthc, 2005, 2, 26–32.

[18] Ozin G. Whatever happened to combinatorial materials discovery? Available at: https://www. advancedsciencenews.com/what-ever-happened-to-combinatorial-materials-discovery/. Accessed February 20, 2020.

[19] Kodadek T. The rise, fall and reinvention of combinatorial chemistry. Chem Commun, 2011, 47, 9757–9763.

[20] Duros V, Grizou J, Xuan W, et al. Human versus robots in the discovery and crystallization of gigantic polyoxometalates. Angew Chemie – Int Ed, 2017, 56, 10815–10820.

[21] Gongora AE, Xu B, Perry W, et al. A bayesian experimental autonomous researcher for mechanical design. Sci Adv, 2020, 6, 1–6.

[22] Shields BJ, Stevens J, Li J, et al. Bayesian reaction optimization as a tool for chemical synthesis. Nature, 2021, 590, 89–96.

[23] Prabhu GRD, Urban PL. Elevating chemistry research with a modern electronics toolkit. Chem Rev, 2020, 120, 9482–9553.

[24] Shao K, Tang Z, Zhu Y, Li N, Zhao D. A survey of deep reinforcement learning in video games. ArXiv, 2019, 1–13.

[25] Vinyals O, Babuschkin I, Czarnecki WM, et al. Grandmaster level in StarCraft II using multi-agent reinforcement learning. Nature, 2019, 575, 350–354.

[26] MacLeod BP, Parlane FGL, Morrissey TD, et al. Self-driving laboratory for accelerated discovery of thin-film materials. Sci Adv, 2020, 6, 1–8.

[27] Flores-Leonar MM, Mejía-Mendoza LM, Aguilar-Granda A, Tribukait H, Amador-Bedolla C, Aspuru-Guzik A. Materials acceleration platforms: On the way to autonomous experimentation. Curr Opin Green Sustain Chem, 2020, 25, 100370.

[28] Häse F, Roch LM, Aspuru-Guzik A. Gryffin: An algorithm for Bayesian optimization for categorical variables informed by physical intuition with applications to chemistry. ArXiv, 2020, 1–32.

[29] Roch M, Häse F, Kreisbeck C, et al. ChemOS: An orchestration software to democratize autonomous discovery. PLoS One, 2020, 15.

[30] Dave A, Mitchell J, Kandasamy K, et al. Autonomous discovery of battery electrolytes with robotic experimentation and machine learning autonomous discovery of battery electrolytes with robotic experimentation and machine learning. Cell Reports Phys Sci, 2020, 2, 100264.

[31] Nikolaev P, Hooper D, Webber F, et al. Autonomy in materials research: A case study in carbon nanotube growth. Npj Comput Mater, 2016, 2, 16031.

[32] Li J, Lim K, Yang H, et al. AI applications through the whole life cycle of material discovery. Matter, 2020, 3, 393–432.

[33] Wang AYT, Murdock RJ, Kauwe SK, et al. Machine learning for materials scientists: An introductory guide towards best practices machine learning for materials scientists: An introductory guide towards best practices. Chem Mater, 2020, 32, 4954–4965.

[34] Cai J, Chu X, Xu K, Li H, Wei J. Machine learning-driven new material discovery. Nanoscale Adv, 2020, 2, 3115–3130.

[35] Chen A, Zhang X, Zhou Z. Machine learning: Accelerating materials development for energy storage and conversion. InfoMat, 2020, 2567–3165.

[36] Himanen L, Geurts A, Foster AS, Rinke P. Data-driven materials science: Status, challenges, and perspectives. Adv Sci, 2019, 6, 1900808.

[37] Severson KA, Attia PM, Jin N, et al. Data-driven prediction of battery cycle life before capacity degradation. Nat Energy, 2019, 4, 383–391.

[38] Vasudevan R, Pilania G, Balachandran P. Machine learning for materials design and discovery. J Appl Phys, 2021, 129, 070401.

[39] Martini A, Guda SA, Guda AA, et al. PyFitit: The software for quantitative analysis of XANES spectra using machine-learning algorithms. Comput Phys Commun, 2020, 250, 107064.

[40] Lee B, Yoon S, Lee JW, et al. Statistical characterization of the morphologies of nanoparticles through machine learning based electron microscopy image analysis. ACS Nano, 2020, 14, 17125–17133.

[41] Ziatdinov M, Dyck O, Maksov A, et al. Deep learning of atomically resolved scanning transmission electron microscopy images: Chemical identification and tracking local transformations. ACS Nano, 2017, 11, 12742–12752.

[42] Ede JM. Deep learning in electron microscopy. Mach Learn Sci Technol, 2021, 2, 013004.

[43] Oviedo F, Ren Z, Sun S, et al. Fast and interpretable classification of small X-ray diffraction datasets using data augmentation and deep neural networks. Npj Comput Mater, 2019, 5, 1–9.

[44] Cutcher-Gershenfeld J, Baker K, Berent N, et al. Five ways consortia can catalyze open science. Nature, 2017, 543, 615–617.

[45] Draxl C, Scheffler M. Big-data-driven Materials Science and Its Fair Data Infrastructure. Handb. Mater. Model., Cham, Springer, 2020, 1–24.

[46] Editorial N. Empty rhetoric over data sharing slows science. Nature, 546, 327.

[47] Coughlan T. The use of open data as a material for learning. Educ Technol Res Dev, 2020, 68, 383–411.

[48] Raccuglia P, Elbert KC, Adler PDF, et al. Machine-learning-assisted materials discovery using failed experiments. Nature, 2016, 533, 73–76.

[49] Wilkinson MD, Dumontier M, Aalbersberg I, et al. The FAIR guiding principles for scientific data management and stewardship. Sci Data, 2016, 3, 1–9.

[50] Pendleton IM, Cattabriga G, Li Z, et al. Experiment specification, capture and laboratory automation technology (ESCALATE): A software pipeline for automated chemical experimentation and data management. MRS Commun, 2019, 9, 846–859.

[51] Soedarmadji E, Stein HS, Suram SK, Guevarra D, Gregoire JM. Tracking materials science data lineage to manage millions of materials experiments and analyses. Npj Comput Mater, 2019, 5, 1–9.

[52] Hessam S, Craven M, Leonov AI, Keenan G, Cronin L. A universal system for digitization and automatic execution of the chemical synthesis literature. Science (80-), 2020, 370, 101–108.

[53] Stein HS, Gregoire JM. Progress and prospects for accelerating materials science with automated and autonomous workflows. Chem Sci, 2019, 10, 9640–9649.

[54] Sanchez-Lengeling B, Aspuru-Guzik A. Inverse molecular design using machine learning: Generative models for matter engineering. Science (80-), 2018, 361, 360–365.

[55] Sun S, Tiihonen A, Oviedo F, et al. A data fusion approach to optimize compositional stability of halide perovskites. Matter, 2021, 4, 1305–1322.

[56] Kalidindi SR. Feature engineering of material structure for AI-based materials knowledge systems. J Appl Phys, 2020, 128, 041103.

[57] Mikolov T, Chen K, Corrado G, Dean J. Efficient estimation of word representations in vector space. 1st Int Conf Learn Represent ICLR 2013 – Work Track Proc, 2013, 1–12.

[58] Tshitoyan V, Dagdelen J, Weston L, et al. Unsupervised word embeddings capture latent knowledge from materials science literature. Nature, 2019, 571, 95–98.

[59] Weston L, Tshitoyan V, Dagdelen J, et al. Named entity recognition and normalization applied to large-scale information extraction from the materials science literature. J Chem Inf Model, 2019, 59, 3692–3702.

[60] Matscholar. Available at: https://matscholar.com/. Accessed April 1, 2021.

[61] Everett L. Lab of the Future. Available at: https://www.labmanager.com/laboratory-technology/lab-of-the-future-431. Accessed March 20, 2021.

[62] Yin H, Fan F, Zhang J, Li H, Lau TF. The importance of domain knowledge. Available at: https://blog.ml.cmu.edu/2020/08/31/1-domain-knowledge/. Accessed March 15, 2021.

[63] Deng C, Ji X, Rainey C, Zhang J, Lu W. Integrating machine learning with human knowledge. IScience, 2020, 23, 101656.

Samira Siahrostami, Stanislav R. Stoyanov, Sergey Gusarov,
Ian D. Gates, Mohammadreza Karamad

2 Artificial intelligence for catalysis

2.1 Introduction

The world's energy systems will need to be changed radically if they are going to supply our energy needs in a sustainable manner over a long-term basis. Access to safe, clean, and sustainable energy supplies is one of the greatest challenges facing humanity during the twenty-first century. Energy from renewable resources – wind, water, the sun, and biomass – is inexhaustible and clean. Solar cells and electrochemical devices such as electrolyzers can be coupled with these intermittent power sources to store electricity in the form of chemical energy, in chemicals such as hydrogen, methanol, ethanol, and ethane (among others) for later use. Such processes have the potential to address the ever-increasing demand for chemicals and fuels while mitigating anthropogenic CO_2 emissions. Catalytic materials play an important role in these promising energy conversion technologies. Based initially on empirical approaches and chemical intuition, catalyst development has been fostered by combining physical and chemical models under the Sabatier's principle, according to which the interactions between the catalyst and the substrate should be neither too strong nor too weak and "just right" [1, 2]. Recent developments of computational tools, such as density functional theory (DFT), have led to an unprecedented understanding of the materials properties and enabled predicting catalytic properties *a priori* [3–8]. Catalyst discovery has been revolutionized as a result of rising computational capabilities and especially by the extensive use of DFT calculations, which provide a powerful computational framework to study catalytic reactions and identify new chemically active materials [9]. In particular, combinatorial quantum-chemical and high-throughput calculations have provided a powerful framework for the discovery of new catalysts [10]. For example, DFT has provided the computational foundation of the hard and soft acid and base (HSAB) principle of Pearson and enabled the calculation of reactivity descriptors, such as hardness, softness, and

Disclaimer: Stanislav R. Stoyanov, © Her Majesty the Queen in Right of Canada, as represented by the Minister of Natural Resources, 2021.
Sergey Gusarov, © Her Majesty the Queen in Right of Canada, as represented by the National Research Council of Canada, 2021.

Samira Siahrostami, Department of Chemistry, University of Calgary, Canada
Stanislav R. Stoyanov, CanmetENERGY Devon, Natural Resources Canada, Canada
Sergey Gusarov, Nanotechnology Research Centre, National Research Council of Canada, Canada
Ian D. Gates, Mohammadreza Karamad, Department of Chemistry, University of Calgary, Canada

https://doi.org/10.1515/9783110738087-002

Fukui functions. The d-band model developed by Nørskov and co-workers to relate the calculated adsorption energy to experimentally measurable catalytic properties.

Despite their promise, DFT calculations are sluggish, and the immense phase space of possible catalytic materials spanned by structural and compositional degrees of freedom hinders rapid catalyst discovery. Therefore, the development of more effective methodologies with acceptable accuracy anchored by DFT calculations is imperative for facile and accelerated catalyst discovery for applications in clean energy technologies, and chemicals and fuels production. One particularly promising approach to address the above challenge is by using machine learning (ML), a subset of artificial intelligence (AI) which has led to a new paradigm shift in research direction for materials discovery [11–21].

The ML models, in particular, facilitate materials discovery by bypassing expensive DFT calculations. The ML models learn trends within materials properties defining the quantities of interest from big materials data. Using ML, promising materials for different applications have been proposed and identified. Although the use of ML for catalyst design goes back to 1990, its application for catalysis has been boosted recently [22–25]. Moreover, despite ML promise in other subfield of materials science, less attention has been paid to developing ML models tailored to catalysis applications [24]. In this chapter, we first describe the developed descriptors of chemical reactivity and then review recently developed ML models for catalysis applications with particular emphasis on the ML models used to predict catalytic activity.

2.2 Prediction of chemical reactivity

The prediction of chemical reactivity has become one of the highest priority tasks of computational chemistry since the development of electronic structure methods [26]. Reactivity prediction generally aims to address the reaction's regioselectivity preference and potential energy surface. The descriptors of chemical reactivity discussed here refer to regioselectivity preferences calculated based on the structure and properties of the reactants. Approaches for relating regioselectivity with the shape of the potential energy surface associated with chemical reactions are also presented. The d-band model proposed specifically for heterogeneous catalysis is discussed with respect to the reactivity descriptors. Reaction preference prediction is closely associated with adsorption and aggregation in the context of interfacial (heterogeneous) and intermolecular (homogeneous) interactions, respectively [27–29]. Adsorption interactions are important in heterogeneous catalysis [30, 31], pollutants removal [32, 33], gas sensors [32], and biomedical applications [34]. Aggregation interactions are highly relevant to supramolecular chemistry [35], asphaltene aggregation [36], and self-assembly in biomolecular systems [37].

2.2.1 Reactivity descriptors

The HSAB principle, proposed by Pearson initially as a qualitative concept, is key to the prediction of chemical reactivity [38, 39]. The acid is the electron acceptor, and the base is the electron donor, as defined by Lewis [40]. Hard acids have high positive charges, small sizes, and low polarizability, whereas soft acids have lower positive charges, larger sizes, and high polarizability. Hard bases have high electronegativity and low polarizability, and are difficult to oxidize, whereas soft bases have low electronegativity and high polarizability and are easily oxidized. The HSAB principle states that hard Lewis acids prefer to react with hard Lewis bases, whereas soft Lewis acids prefer to react with soft Lewis bases in terms of both their kinetic and thermodynamic properties [38, 39]. Nearly three decades later, Pearson and co-workers formulated the chemical potential (μ) and chemical hardness (η) as the first and second derivatives of the ground state electronic energy $E_v[\rho]$ with respect to the number of electrons (N), respectively, under the constant external potential $v(\vec{r})$ that also includes the nuclear potential [41–43]:

$$\mu = \left(\frac{\partial E_v[\rho]}{\partial N}\right)_{v(\vec{r})} = \left(\frac{\delta E_v[\rho]}{\delta \rho}\right)_{v(\vec{r})} = -\chi \tag{2.1}$$

$$\eta = \left(\frac{\partial^2 E_v[\rho]}{\partial N^2}\right)_{v(\vec{r})} = \left(\frac{\partial \mu}{\partial N}\right)_{v(\vec{r})} \tag{2.2}$$

In the second part of equation (2.1), the chemical potential is expressed based on the functional relationship of the energy to the electron density (ρ). The chemical potential is equal to the negative of the electronegativity (χ), as shown in the last part of equation (2.1). These definitions represent an important fundamental connection between HSAB and DFT, as in the latter the properties of a chemical system are expressed as functions of the electron density, and thus establish the quantum chemical foundations for the prediction of chemical reactivity. Also, the chemical softness (σ) is defined as the reciprocal of the chemical hardness:

$$\sigma = \frac{1}{\eta} = \left(\frac{\partial^2 N}{\partial E^2}\right)_{v(\vec{r})} = \left(\frac{\partial N}{\partial \mu}\right)_{v(\vec{r})} \tag{2.3}$$

A chemical reaction, in general, is governed by the principle of electronegativity equalization [44], which assumes the balancing of chemical potentials (equation (2.1)) of reactants through the redistribution of electronic density. According to the HSAB principle, a reaction between an acid A and a base B would be favored when the global softness difference $\Delta\sigma$, defined in equation (2.4), is minimal [45, 46]. This is obtained through the optimization of the covalent contribution of the interaction energy, consequently neglecting other effects, such as polarization [47]

$$\Delta\sigma = \sigma_A - \sigma_B \tag{2.4}$$

These quantitative acid–base reactivity criteria that pertain to the entire molecule, referred to as global, are used to determine the reactivity preferences according to the HSAB principle. In practical calculations, the chemical potential, hardness, and softness (equations (2.1)–(2.3)) are computed by using the finite difference approximation:

$$\mu = -\frac{IP + EA}{2} = \frac{E_{HOMO} + E_{LUMO}}{2} \tag{2.5}$$

$$\eta = IP - EA = \frac{E_{LUMO} - E_{HOMO}}{2} \tag{2.6}$$

$$\sigma = \frac{1}{IP - EA}, \tag{2.7}$$

where IP and EA stand for the vertical ionization energy and the vertical electron affinity, respectively, of the N-electron system. Also, $E_{\{HOMO\}}$ and $E_{\{LUMO\}}$ are the energies of the highest occupied and lowest unoccupied molecular orbitals, respectively, of the same N-electron system [42, 47]. Alternative ways of approximating these energy derivatives have been discussed in depth elsewhere [48, 49].

The reactivity of particular sites in a molecule can be predicted by using local descriptors, such as electron density ($\rho(\vec{r})$) and Fukui function ($f(\vec{r})$), that are dependent on the reaction coordinate. The main variable in DFT and the fundamental site reactivity descriptor is the electron density [50]:

$$\rho(\vec{r}) = \left(\frac{\delta E_v[\rho]}{\delta v(\vec{r})}\right)_N \tag{2.8}$$

The Fukui functions were introduced in 1952–1954 by Fukui and coworkers as an approach to predict the sites of electrophilic, nucleophilic, and radical attacks based on frontier orbitals [51, 52]. Decades later, Parr and Yang pointed to the strong connection of the Fukui functions to chemical reactivity prediction using DFT [42]. The Fukui function is defined in DFT as the derivative of the electron density at a position $\vec{r}$ in the three-dimensional (3D) space $\rho(\vec{r})$ with respect to the number of electrons of the system N under a constant external potential $v(\vec{r})$. Alternatively, the Fukui functions are expressed as a variational derivative of the chemical potential μ with respect to the external potential $v(\vec{r})$ at a constant number of electrons N [42]:

$$f(\vec{r}) = \left(\frac{\partial \rho(\vec{r})}{\partial N}\right)_{v(\vec{r})} = \left(\frac{\delta \mu}{\delta v(\vec{r})}\right)_N \tag{2.9}$$

From the last part of equation (2.9), one can see that the Fukui function expresses the sensitivity of the chemical potential of a quantum system to an external perturbation represented by $\delta v(\vec{r})$.

The nucleophilic, electrophilic, and radical Fukui functions are defined in equations (2.10), (2.11), and (2.12), respectively [42]:

$$f^+(\vec{r}) = \left(\frac{\partial \rho(\vec{r})}{\partial N}\right)^+_{v(\vec{r})} = \rho_{N+1}(\vec{r}) - \rho_N(\vec{r}) \tag{2.10}$$

$$f^-(\vec{r}) = \left(\frac{\partial \rho(\vec{r})}{\partial N}\right)^-_{v(\vec{r})} = \rho_N(\vec{r}) - \rho_{N-1}(\vec{r}) \tag{2.11}$$

$$f^0(\vec{r}) = \frac{f^+(\vec{r}) + f^-(\vec{r})}{2} = \rho_{N+1}(\vec{r}) - \rho_{N-1}(\vec{r}) \tag{2.12}$$

where the superscripts + and − denote that the derivatives are taken from above and below, respectively, and the superscript 0 refers to a radical. As the number of electrons N is a discrete variable, the right and the left derivatives of $\rho(\vec{r})$ with respect to N are discontinuous. The parameters on the right side of equations (2.10)–(2.12) are calculated by using the finite difference approximation, where the subscripts $N+1$, N, and $N-1$ refer to the cation, neutral, and anion, respectively. It is important to note that the electron densities $\rho_{N+1}(\vec{r})$ and $\rho_{N-1}(\vec{r})$ are calculated at the optimized geometry in the ground state. The Fukui functions can also be calculated by using the finite difference approximation for non-integer charge increments, e.g., $N+0.1$, N, and $N-0.1$ [53]. The finite difference approximation to the Fukui function is exact at the zero-temperature limit [47, 54]. While there are similarities between the interpretation of the Fukui functions and the frontier molecular orbital theory proposed by Fukui [55, 56], it is important to note that Fukui functions incorporate electron correlation and orbital relaxation, effectively superseding the initial molecular orbital basis [42, 57]. Moreover, the Fukui function is a response function and carries predictive information of the expected change of the electronic density of a molecular system to the addition or removal of a given number of electrons.

Electrons tend to move from reactants with high chemical potential to reactants with low chemical potential to balance the resulting value of the chemical potential. The Fukui function characterizes the local ability of a molecular system to donate and accept electrons. For example, the regions with large positive values of f^+ are the most favorable for a nucleophilic attack, while the regions with large positive values of f^- are the most suitable for electrophilic attack. The interplay of nucleophilic and electrophilic Fukui functions is defined by the ratio of the chemical potentials of the reactants [58]. In case of a large difference in the chemical potentials of reactants, the reactant with a higher chemical potential would transfer a number of electrons ($\Delta N > 0$) from regions, where its f^- is large, to the regions of the reactant with a lower chemical potential, where f^+ is large. If the values of chemical potentials are similar, the reactants equally donate and accept electrons, and a full derivative (equation (2.12)) is needed to describe the density redistribution. An approach for calculating Fukui functions as the functional derivative of the chemical potential with respect to the external potential (second part of equation (2.9)) has also been suggested [59].

Figures 2.1 and 2.2 display the application of the Fukui functions for the prediction of intermolecular aggregation and molecule–surface adsorption preferences, respectively, in an example illustrating the regioselectivity preference analysis of the vanadyl porphyrin and pyridine molecules, representative of petroleum asphaltene moieties. The analysis, conducted using the nucleophilic and electrophilic Fukui functions (Fig. 2.1), shows these amphiphilic molecules can interact in two distinct modes. In Fig. 2.1(a), pyridine is considered an acid and vanadyl porphyrin is considered a base. The maximum of the nucleophilic Fukui function (f^+) of pyridine, shown in red color, is localized in the area near the H atom on the para-position to the N atom of pyridine. The maximum of the electrophilic Fukui function (f^-) of vanadyl porphyrin is localized on the O atom of vanadyl (area shown in red color). The product of this acid–base reaction would be a hydrogen-bonded aggregate. In Fig. 2.1(b), pyridine and vanadyl porphyrin are considered as a base and an acid, respectively. The maximum of the electrophilic Fukui function of pyridine (f^-), which is the preferred site for an attack by an electrophile, is localized on the N atom (area shown in red color). The maximum of the nucleophilic Fukui function (f^+) of vanadyl porphyrin, which is the preferred site for an attack by a nucleophile, is localized on the V atom (area shown in red color). The product of this acid–base reaction is an octahedral complex containing pyridine axially coordinated to vanadyl porphyrin. The predicted axial coordination and hydrogen bonding aggregation modes yield stable products, as confirmed by full geometry optimization using the

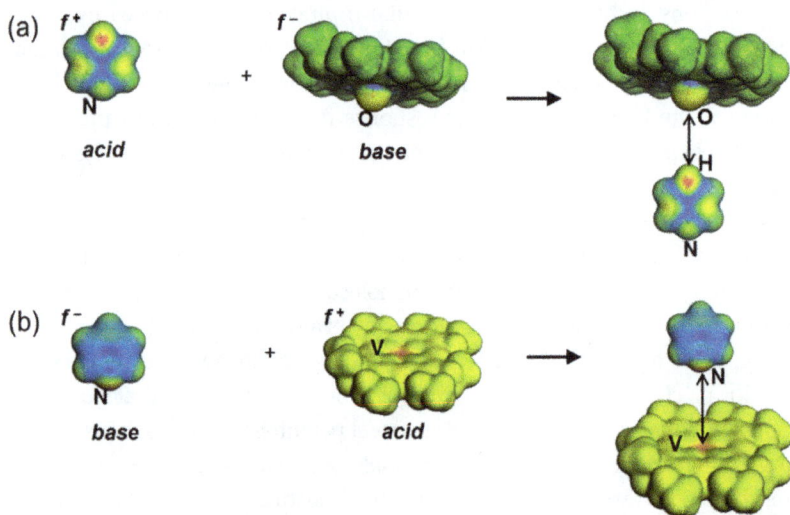

Fig. 2.1: Regioselectivity preferences between vanadyl porphyrin and pyridine by using nucleophilic and electrophilic Fukui functions (mapped on the 0.017 e/Å3 electron density isosurface), calculated by using DMol3/PBE/DNP: (a) Pyridine and vanadyl porphyrin as acid and base, respectively; (b) Pyridine and vanadyl porphyrin as base and acid, respectively. The f^- and f^+ minima to maxima values are mapped in blue to red colors, respectively. Adapted from [60] © 2021 Her Majesty the Queen in Right of Canada.

DMol^3 software with the PBE0 functional and DNP basis set [60]. These aggregation interactions are accounted in the supramolecular assembly model [36]. Highly robust and persistent over geological time periods, vanadyl porphyrins are found in petroleum asphaltenes and thought to be derived from Mg(II) porphyrins in chlorophylls during diagenesis, as vanadyl porphyrins are inert [61]. Vanadium recovery from petroleum asphaltenes is considered a step of an integrated approach for extracting more value from Canada's oil sands [62]. Nitrogen-containing ligands could be tethered to help form vanadyl porphyrin clusters, suggesting an approach for vanadium recovery [63]. Vanadium is increasingly important for energy storage as the main component of vanadium flow batteries that outperform Li-ion batteries in long-duration applications, where energy is required daily [64].

Figure 2.2 displays the predicted adsorption configuration of benzene on Cu+ exchanged chabazite, a porous zeolite catalyst. The red region at the Cu atom indicates the maximum of f^+, which is the preferred site for a nucleophilic attack. The yellow region of benzene highlights the maximum of f^-, which is delocalized over the π-electron system. The preferred adsorption interaction (highlighted in gray) involves the coordination of a C=C site of benzene with the Cu atom of chabazite, as confirmed by a full geometry optimization using DMol^3 with the PW91 functional and DNP basis set [65]. Fukui functions can be employed for the rational design of catalytic zeolite nanoparticles [75–77].

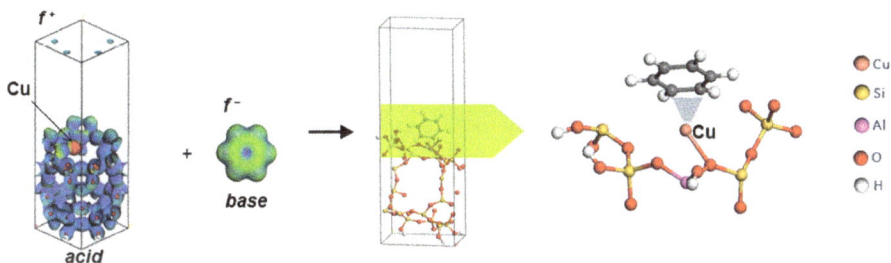

Fig. 2.2: Nucleophilic Fukui function of Cu$^+$ exchanged chabazite slab and electrophilic Fukui functions of benzene mapped on the 0.02 e/A^3 electron density surface are used to predict the adsorption configuration, enlarged (right) to show the bonding mode highlighted in gray (calculated by using DMol^3/PW91/DNP). The f^- and f^+ minima to maxima values are mapped in blue to red colors, respectively. © 2021 Her Majesty the Queen in Right of Canada.

Chemical reactions have been considered as being controlled by electrostatic (charge) or frontier orbital (electron transfer) effects [66]. Frontier orbital-controlled reactions typically feature neutral molecules with slightly charged active sites, corresponding to soft-soft interactions and the breaking and formation of covalent bonds. Electrostatically controlled reactions involve charged species (hard acids and bases) and ionic bond formation and breaking [67]. However, there is evidence of a continuum of cases of chemical reactions between amphiphilic reagents where both frontier orbital and electrostatic effects are important [66, 67]. Reactivity descriptors that aim to address the dualistic nature of amphiphilic reagents have been proposed [68]. A dual descriptor for nucleophilicity and electrophilicity has been introduced as the second derivative of the electron density with respect to the number of electrons [69]:

$$\Delta f(\vec{r}) = \left(\frac{\partial^2 \rho(\vec{r})}{\partial N^2}\right)_{v(\vec{r})} \simeq \frac{f^+(\vec{r}) - f^-(\vec{r})}{2} \qquad (2.13)$$

This dual descriptor, defined as the difference between the nucleophilic and electrophilic Fukui functions, has been employed to predict stereoselectivity in dual ion-molecule complexes and substituted phenyls [68, 69]. It is important to note that the orbital-based dual descriptor has to be used in conjunction with the electrostatic potential [68]. Another comprehensive descriptor is the Fukui potential, defined as [67]:

$$v_f^{+/-}(\vec{r}) = \int \frac{f^{+/-}\left(\vec{r'}\right)}{\left|\vec{r} - \vec{r'}\right|} d\vec{r'} \qquad (2.14)$$

The Fukui potential approximates the molecular hardness distribution and determines the electron transfer active site, which makes it the electrostatic contribution to the frontier local hardness. The shape of the Fukui potential guides an approaching reagent toward the reactivity site, where an electrophilic or nucleophilic site prefers to accept or donate charge, respectively, in the course of a reaction. At the position of the nuclei, the Fukui potential is equal to the variation of the chemical potential with the nuclear charge and therefore measures the sensitivity of the system to changes in atom type, which is particularly useful for designing molecular structures with desirable electronic properties [67].

Atomic or site descriptors of reactivity, such as atomic charge, condensed Fukui function, local softness, local hardness, and local philicity are employed to quantify the chemical reactivity of atoms and functional groups within molecules. The atomic charge is an important and widely used reactivity descriptor, in particular, for electrostatically controlled reactions between hard acids and bases. The predictive capacity of charge analysis methods based on fitting to the molecular electrostatic potential, partitioning in terms of atomic orbitals, and atoms-in-molecules partitioning of the electron density has been discussed elsewhere [70, 71].

In the finite difference approximation, the condensed or atomic Fukui functions for nucleophilic and electrophilic attacks on atom x in a molecule that contains N electrons are defined in equations (2.15) and (2.16), respectively [47]:

$$f_x^+ = q_x(N+1) - q_x(N) \tag{2.15}$$

$$f_x^- = q_x(N) - q_x(N-1) \tag{2.16}$$

The nucleophilic and electrophilic local softness σ_x^+ and σ_x^- for atom x are the product of the condensed nucleophilic and electrophilic Fukui functions f_x^+ and f_x^-, respectively, and the global softness σ, as shown in equations (2.17) and (2.18) [72]. The local softness contains the same local reactivity information as the condensed Fukui function plus additional information about the total molecular softness, which is related to the global reactivity with respect to a reaction partner

$$\sigma_x^+ = (f_x^+)\sigma \tag{2.17}$$

$$\sigma_x^- = (f_x^-)\sigma \tag{2.18}$$

The relative nucleophilicity and relative electrophilicity are defined as (σ_x^+/σ_x^-) and (σ_x^-/σ_x^+), respectively. Both Brønsted and Lewis acidity have been characterized by using nucleophilicity and electrophilicity indices, respectively [73, 74]. Detailed discussions on reactivity indices are available elsewhere [75, 76]. It is important to note that the atomic partitioning of the condensed Fukui function should be conducted with caution due to challenges associated with widely used population analysis methods. The partitioning using Mulliken population analysis for computing condensed Fukui functions leads to uncertainty, as the weight factor on the electron density distribution is dependent on whether the system is neutral, cationic, or anionic [77]. Moreover, an alternative, iterative version of the Hirshfeld population analysis procedure has been proposed to address partitioning challenges arising from the choice of the promolecular density [78].

The descriptors of chemical reactivity can be extended to periodic systems such as metals, a case that is of particular importance for heterogeneous catalysis. At T = 0 K, the metal chemical potential is equal to the Fermi energy ε_f and the global softness (reciprocal of the global hardness) is equal to the density of states $g(\varepsilon_f)$ [79, 80]:

$$\sigma = \frac{1}{\eta} = \left(\frac{\partial N}{\partial \mu}\right)_{T,V} = g(\varepsilon_f) \tag{2.19}$$

where V is the volume. The lattice structure remains unchanged in the differentiation. The Fukui function is the normalized local density of states at the Fermi energy:

$$f(\vec{r}) = \frac{g(\varepsilon_f, \vec{r})}{g(\varepsilon_f)} \tag{2.20}$$

where $g(\varepsilon_f, \vec{r})$ is the local density of states:

$$g(\varepsilon, \vec{r}) = \sum_i |\psi_i(\vec{r})|^2 \delta(\varepsilon_i - \varepsilon) = \frac{2V}{(2\pi)^3} \int d\vec{k} |\psi_k(\vec{r})|^2 \delta(\varepsilon(\vec{k}) - \varepsilon), \qquad (2.21)$$

where $\psi_i(\vec{r})$ are the normalized Kohn–Sham orbitals. These definitions enable the description of molecules and periodically extended systems in a unified way. Equation (2.20) allows the association of the Fukui function with the reactivity of metals. In general, metals are soft reagents [81] with large $g(\varepsilon_f)$ and transition metals are particularly active because of their large density of states. Therefore, adsorption and catalytic reactions on transition metal surfaces are considered soft–soft chemical reactions [79].

The calculations of reactivity descriptors based on the ground state energy and electron density, such as hardness, softness, and especially Fukui functions faces challenges for systems with degenerate orbitals and states. While in principle, conceptual DFT is exact, most of its applications employ approximate density functionals based on independent electron descriptions. The single-determinant approximation used in Kohn–Sham DFT produces inaccurate reaction barriers and excited states that are important in the cases of complex electronic structure redistribution, as in Fukui function calculations. Degenerate frontier orbitals and states as well as electron correlation effects also influence the calculated reactivity descriptors. The rigorous investigations of conceptual DFT require the implementation of more accurate computational methods. To address these computational needs and enable more accurate calculations of Fukui functions, approaches based on the concept of the Koopmans' theorem extended for the multiconfigurational case have been proposed. The extended Koopmans' theorem describes the change of the electron density due to the removal or addition of an electron, which relates to the Fukui function (equations (2.10)–(2.12)). Ayers and Melin have proposed approaches based on the differential equation of the extended Koopmans' theorem as well as its ionized wave function expression. They report that for electron removal processes the extended Koopmans' theorem is superior to the electron propagator theory. However, they note shortcomings when the electron addition is computed and propose to start from the $N+1$ electron system [82]. Gusarov et al. have proposed methodologies to construct Green's functions within the Koopman's approximation for the multiconfigurational case [26, 83]. The initial formulation is based on the concept of the partial summation of perturbation series that leads to a replacement of initial electron propagators by propagators in the multiconfigurational approximation [83, 84]. Subsequently, the methodology has been improved by considering a partial summation of the perturbation expansions for the one-particle Green's function. Alternative approximations of the multiconfigurational Green's function have been considered in case of a large dimension of the reference configuration space [26]. Recently, the extended Koopmans' theorem has been combined with the second-order Møller–Plesset perturbation theory and the adiabatic connection formalism of DFT to address degenerate orbital occupation challenges [85, 86]. An orbital-weighed Fukui function based on the dual descriptor (equation (2.13)) in terms of Koopmans' approximation has been proposed for regioselectivity prediction in systems with (quasi-)degenerate frontier molecular orbitals [87].

The Fukui matrix has been introduced as an approach to analyze the Fukui function in systems with degenerate frontier orbitals. Diagonalizing the Fukui matrix gives a set of eigenvectors, the Fukui orbitals, and accompanying eigenvalues. The dominant eigenvector is used to evaluate the frontier molecular orbital picture quality, especially for systems containing degeneracies [88, 89].

Approaches based on gradient bundle volumes and valence band theory have also been proposed for regioselectivity prediction. For example, Morgenstern et al. have proposed a method to predict the most probable regions for electrophilic attack within each atom by using gradient bundle volumes, properties that depends only on the charge density of the neutral molecules [90]. They introduced gradient bundle condensed Fukui functions to compare the stereoselectivity information obtained from gradient bundle volume analysis and demonstrate this method using a test set of molecular fluorine, oxygen, nitrogen, carbon monoxide, and hydrogen cyanide. The qualitative valence bond analysis has been employed to demonstrate the relationship between regioselectivity preferences based on local reactivity descriptors and the potential energy surface of chemical reactions. The valence bond theory aims to predict the impact of individual local reactivity descriptors on the global potential energy surface based on the inherent connection between spatial and energetic stabilization. The effort to unify the valence bond theory with conceptual DFT is ongoing [91].

2.2.2 The *d*-band model

Hammer and Nørskov have proposed a methodology to relate the calculated chemisorption of reaction intermediates to experimentally observed catalytic behavior [92–96]. They have formulated the *d*-band model to relate chemisorption energies to a single state with energy ε_d corresponding to the center of the *d*-band, which can be calculated and measured. The *d*-band center (ε_d) is defined as:

$$\varepsilon_d = \frac{\int \varepsilon g(\varepsilon, \vec{r}) d\varepsilon}{\int g(\varepsilon, \vec{r}) d\varepsilon} \tag{2.22}$$

where ε is the energy and $g(\varepsilon, \vec{r})$ is the density of states at energy ε and position $\vec{r}$ [97, 98]. Using the *d*-band center enables the mapping of parameters that determine the rate of the catalytic reaction onto a reduced space spanned by energy parameters, known as descriptors. The change of the adsorption energy is then correlated with a shift in ε_d. A stronger upward shift indicates the possibility of the formation of a larger number of unoccupied anti-bonding states, leading to stronger bonding. Therefore, the position of the *d*-band center can be considered an indicator or descriptor of catalytic reactivity. The Hammer–Nørskov model successfully explains both the experimental and first principle theoretical results for different heterogeneous catalytic processes. In subsequent studies, the position of the d-band center was shown to correlate with the adsorption, activation, and dissociation energies of small molecules, such as N_2 and CO, providing a simple and efficient precursor to many chemical processes [99, 100]. This model has successfully been generalized to explain the catalytic activity of magnetically polarized transition metal surfaces [101]. It has also been confirmed in numerous experimental and theoretical reports that the modeling of the adsorption of small molecules can give insights that are difficult to obtain from experiments alone and are valuable for sustainable chemical production, alternative energy solutions, and pollution mitigation [99, 102].

Several studies relate the d-band theory with the descriptors of chemical reactivity within the HSAB theory [80, 103]. The predictive capacity of the Fukui functions of Au, Pt, Pt, with Au core and Pt with Ru core nanoparticles has been evaluated with respect to the d-band center values and correlated with the CO adsorption energy. The results have shown a good comparison between Fukui functions and d-band center in that as the d-band center becomes more positive, the Fukui function values grow larger. Moreover, the hardness and softness are found to correlate well with the Fermi-level density of states and CO binding energy. Challenges related to the use of the dual descriptor (equation (2.13)) have been attributed to numerical issues arising from the degenerate states and orbitals [80]. The adsorption energy of the O atom on Pt, Pd, Rh, and Au clusters supported on metal oxides has been expressed in terms of the d-band centers, chemical potentials, hardness, and softness (defined in equations (2.1)–(2.3) and (2.22)) and interpreted in terms of the HSAB theory [103]. Fukui function and d-band center values have been employed to correlate reactivity with coordination number of Au–Pd nanoparticles, with the two descriptors being in a good agreement [104]. Gao et al. proposed an electronic descriptor that effectively reflects the d-band characteristics of metallic materials and valence of surface atoms in oxides and employed it to in a model to determine the adsorption energies of reaction intermediates on a series of surfaces and nanoparticles [105].

2.3 Machine learning models for chemical reactivity prediction

The ML models used in catalysis mainly consist of three parts, as depicted in Fig. 2.3: generating datasets including a relatively large amount of computational or experimental data on specific materials relevant to the problem, numerical representation of the catalysts (input features) [106], and an ML algorithm that needs to be compatible with the nature of the data and the numerical representation. The ML algorithm is intended to train the ML model by building a functional map between the input data and output data being the catalytic properties of interest such as binding energies of adsorbates.

During training, the ML model learns the underlying connection between the catalyst properties and the quantities of interest [107]. At this stage, the ML model can be used to either screen new catalyst materials and predict their quantities of interest with the purpose of finding optimal catalysts or interpret catalyst figures of merit from input features.

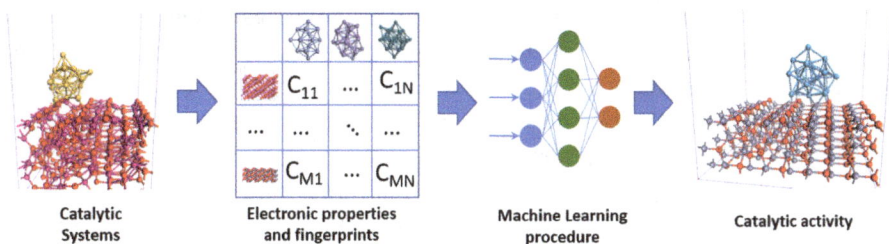

| Catalytic Systems | Electronic properties and fingerprints | Machine Learning procedure | Catalytic activity |

Fig. 2.3: ML procedure workflow: System selection, calculation of electronic properties, use of molecular fingerprints, application of a trained ML procedure, and identification of the best system. Prepared using BIOVIA Discovery Studio 2021, Dassault Systèmes BIOVIA, D. S. M. E., Release 2021, San Diego: Dassault Systèmes, 2021. © 2021 Her Majesty the Queen in Right of Canada.

In the rest of this chapter, we review recently developed ML models for catalysis applications with a particular emphasis on those used to predict catalytic activity. We only focus on supervised ML models for catalysis applications where the machine learning model maps out input data to targeted properties. We start by discussing the required components for the development of a supervised ML model for catalysis.

2.4 Catalysis databases

Databases that include a relatively large amount of computational or experimental data or both on specific materials relevant to the problem are essential ingredients for the development of viable ML models. The effect of databases on the development of viable ML models for catalysis is two-fold: First, the size and nature of a database is directly related to the performance of an ML model for predicting the desired catalytic properties of new materials. For example, an ML model developed using a database containing a limited chemical space of existing materials is expected to exhibit low performance when used to predict the targeted properties of materials belonging to a totally different chemical space. Second, the size of the database is related to the ML models, as the database size used to train an ML model could determine the model's performance.

Large libraries of materials data, such as the materials project (MP) [108–111], open quantum materials database (OQMD) [112], and computational materials repository (CMR) [113] have been made available in recent years. However, their application is limited to investigating bulk properties, such as formation energies, band structure, and density of states. On the other hand, surface catalytic properties which are vital for describing the chemical reactivity of solid surfaces cannot be directly captured using bulk properties. As a result, these databases cannot be directly used for the development of ML models relevant to heterogeneous catalysis. Accurate and computationally expensive quantum-chemical calculations are required to generate databases for catalyst materials that are not limited to small subsets of materials space with a narrow range of catalytic properties. This urgent matter has gained momentum in recent years and led to the development of comprehensive databases, such as Open Catalyst 2020 (OC20), consisting of 1,281,040 DFT calculations spanning a wide range of materials, surfaces, and adsorbates [114]. Another open-source catalysis database is the Catalysis Hub, containing thousands of reaction energies and barriers from DFT calculations on heterogeneous systems [115, 116]. However, there is still an urgent need to develop databases that cover other classes of materials such as oxides that have a wide range of applications in chemical and electrochemical processes related to clean energy technologies.

2.5 Features development and materials representation

Another important step in the development of a robust ML-based approach to predict the properties of catalytic materials is the design or selection of features (to be input in the ML model) that can accurately represent the local chemical environment of atoms on the catalyst surface. The features need to have three important characteristics: 1) they must be unique in representing a material in order to enable the ML model to distinguish materials from each other, 2) they should preferably be calculated at a low computational cost or preferably be readily accessible from available databases, and 3) they should reflect the chemical information related to the targeted properties. Developing features that possess the aforementioned characteristics has proven to be challenging [101]. In this section, we first briefly discuss different materials representation methods that have been used for crystals in materials science community. Then, we discuss different featurization methods that have been developed for heterogeneous catalysis applications.

2.5.1 Materials featurization for crystals

For crystals, different types of features have been developed for materials representation in the past [117, 118]. Examples include the smooth overlap of atomic positions that encodes the local chemical environment of a center atom as smoothed Gaussian densities of its neighboring atoms [119]. The many-body tensor representation encodes the periodic crystalline structures as a whole by expanding them in a distribution of different structural motifs that are based on chemical elements [120]. The Coulomb matrix representation captures the atomic interactions based on nuclear charges and atomic positions [121]. These featurization methods benefit either from structural or elemental information or both. Elemental features often include intrinsic quantities such as atomic number or heuristic quantities, such as electronegativity and ionic radius that can be readily accessed [122–124]. Structural representations, on the other hand, encode local chemical environment of atoms by capturing the geometry and interactions among atoms [125].

Another important type of features in defining materials properties are the electronic structure features [14, 122]. Intuitively, their inclusion as features when encoding the local chemical environment of atoms is expected to increase the performance of an ML model. Examples of such attributes are the density of states (DOS), characteristics of band structures, Fukui functions (Section 6.1) and orbital–orbital interactions [126]. The explicit inclusion of these features, however, is associated with extra computational costs. Meanwhile, efforts have been made to include some aspects of electronic structure attributes without performing DFT calculations. For example, Ward et al. used the average fraction of electrons from the s, p, d, and f valence shells among all present elements as electronic structure attributes [123]. In another effort, Pham et al. developed a two-dimensional descriptor called the orbital field matrix to encode orbital interactions by considering the electron configurations of each atom and its neighbors in the crystal [126]. Some of features that were originally developed for crystals have been adopted for catalysis applications [118, 127, 128].

2.5.2 Materials featurization for heterogeneous catalysis

For catalytic applications, the figures of merit such as catalytic activity, selectivity and stability are defined through the interaction of adsorbates with the catalyst's surface at atomic scale. Therefore, in developing features for predicting catalytic properties, it is important to account for features that govern the interaction of adsorbate with catalyst surface. It has been shown that electronic structure and geometrical attributes play important roles in defining the binding energies of adsorbates. This, in turn, implies that the accuracy of a ML model to predict material properties is mostly controlled by the ability of its descriptors to account for such attributes and therefore accurately encode the local chemical environment of atoms at the catalyst surface. The d-band approach is the most relevant electronic structure attribute model that has been proven to be useful in describing both the bond formation at transition metal surfaces and the trends in transition metal reactivity. Therefore, the characteristics of the d-states yield useful electronic structure features that can been used for materials representation when developing ML models for catalytic properties prediction. Although valuable, unfortunately, the inclusion of d-band characteristics as features requires extra DFT calculations. The latter, in particular, becomes computationally prohibitively expensive when one needs to consider the *d*-band characteristics of multiple surface sites on a catalyst surface. Noh et al. have proposed the use of *d*-band width within the muffin-tin orbital theory as features requiring no further DFT calculations [129]. In another approach, efforts have been made to predict the d-band center using basic atomic properties [130]. Such ML approaches, however, may not be generalizable to other classes of materials in the catalysis space for predicting *d*-band characteristics and their use in a subsequent ML model for property prediction may be limited. For example, the metal d-bands of metal oxides are different from those of pure transition metals and separate reactivity descriptors have been proposed for each [131, 132].

Previous studies have shown that structural attributes such as coordination number or orbital-wise coordination number are powerful descriptors for trends in binding energies of adsorbates. Inspired from these studies, coordination numbers have been used as features to represent catalyst materials [133–135]. In addition, other types of features that have been used to represent crystals in ML models for catalytic properties predictions take into account the atomic properties of their constituent elements. Considering their simplicity and low computational costs compared to extracted features, these structural attribute-based features have shown great promise for catalytic properties prediction. For example, using only geometrical, elemental, and a modified form of coordination numbers as features for predicting the binding energy of CO on a wide range of metallic alloys, have been reported to achieve a higher accuracy than computationally expensive *d*-band characteristics.

In the above featurization methods, the features are constructed manually. The disadvantage of manual featurization, often called feature engineering, is that only a limited subset of aspects of the materials is considered, while others are ignored. It would be desirable, if the model itself could derive features and make decision based on the input data. To address this challenge, the integration of features into a deep neural networks model, such as convolutional neural networks (CNNs) has been widely used where the models have a strong ability to extract complex relevant features that are impossible otherwise to design manually [136]. Another commonly used representation in conjunction with CNNs for feature extraction is the graph representation [127, 137–141]. By using CNNs on the top of the graph representation, different ML models for predicting the properties of both crystals and catalysts have been developed [137–143]. Some of these models will be discussed in the ML models section below.

2.6 ML models

The supervised ML models can be categorized as classification and regression, where the former predicts a discrete class label output and the latter predicts a continuous quantity output. Based upon their algorithms, the regression models are further categorized as analytical and neural network or deep learning. The analytical models, on one hand, offer higher interpretability than neural network and deep learning models when used for predicting catalytic properties [144]. Neural networks and, in particular, deep learning models, on the other hand, offer higher predictive accuracy than analytic models. In other words, the high accuracy comes at the cost of lower interpretability. Other ML models that take advantage of statistical inference, such as active learning, [145], allow the navigation of large materials space for guiding the computations through uncertainty quantification [146]. Particular attention has been paid to using the above ML models for predicting the chemical reactivity of catalysts relevant to different chemical and electrochemical reactions. More details about different ML models for catalysis application can be found in recently published reviews [24, 101, 142]. Later in this chapter, we describe some of the ML models along with their respective featurization methods for catalysis application in electrochemical reactions, such as CO_2 reduction reaction (CO_2RR), oxygen evolution reaction (OER), and methanol electro-oxidation.

2.6.1 CO_2 reduction reaction

The CO_2RR using renewable electricity, also known as artificial photosynthesis, has emerged as an appealing process for mitigating the effect of CO_2 emissions and converting this greenhouse gas to valuable chemicals, thereby closing the carbon cycle. Transition metals and, in particular, copper, have been widely studied both experimentally and computationally as catalysts for CO_2RR. However, the efficiency of pure metals as CO_2RR catalysts is still lower than required for commercial applications. This calls for the development of new catalytic materials that offer higher efficiency. Other classes of materials, such as bimetallic alloys, are potentially attractive because these offer larger compositional and structural degrees of freedom, hence new active sites. Despite their promise, it has been challenging to identify desirable active site motifs on alloys. Limited computational resources hinder the high-throughput screening of the diverse chemical space of alloys to identify promising catalysts [100, 147–150]. In recent years, different ML models have been developed to efficiently address this challenge [19, 24, 144, 151–154]. Several artificial neural network models have been used to screen and predict the catalytic activities of large libraries of alloys containing both large composition spaces and complex active sites. The NiGa bimetallic alloy is a good example that has been studied to identify active site motifs in a wide range of surface terminations using DFT calculations in parallel with neural network models [100]. In this study, the binding energy of CO has been used as a descriptor of catalytic activity for CO_2RR [155]. Different Ni–Ga compositions, including NiGa, Ni_3Ga, and Ni_5Ga_3, have been considered to make 40 surface terminations and 583 unique adsorption sites (Fig. 2.4(a)). To develop features that uniquely describe adsorption sites, seven different adsorption site characteristics defined based on the coordination numbers have been employed. In addition, only atoms

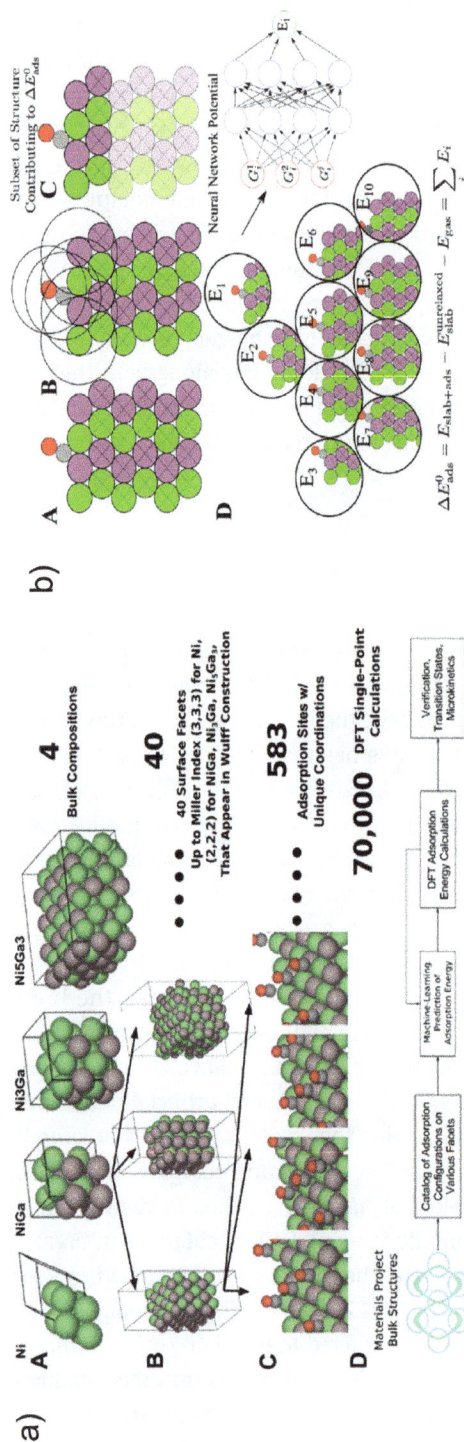

Fig. 2.4: (a) Challenges to the identification of active sites and surfaces of bimetallic catalysts: (A) 4 different catalyst compositions; (B) 40 different facets/terminations for Miller indices up to (333); (C) 583 adsorption configurations identified on all facets/terminations with unique average coordination of bonding metal atoms; (D) High-throughput scheme developed for screening. (b) Schematic of the neural network potential used for predicting the binding energies of small molecules, such as CO: (A) A typical bimetallic surface with a CO adsorbate, with the top layer free to relax; (B) Atoms near those with degrees freedom are included in the reduced representation; (C) Structure subset used to predict the adsorption energy; (D) Local regions used to generate geometric fingerprints, which are analyzed through a neural network to provide atomic contributions to the adsorption energy. The predicted adsorption energy is a summation over these atomic contributions. Adapted from [100] with permission from the American Chemical Society.

on the catalysts surfaces and the neighboring atoms within 3.5 Å of constrained atoms (the atoms that are not fixed and allowed to relax) have been taken into account. To predict the binding energy, a surrogate model using per-atom neural network potentials has been constructed (Fig. 2.4(b)) [160]. The per-atom neural network potentials have enabled the prediction of binding energies with only two single-point DFT calculations. In this work, several Ni–Ga facets with optimal CO binding energy including NiGa(210), NiGa(110), and $Ni_5Ga_3(021)$ have been identified. All of these facets have been found to contain undercoordinated exposed Ni sites surrounded by Ga atoms with no nearby Ni atoms. The identified active site motifs corresponding to facets with optimal catalytic activity for CO_2RR have been employed to rationalize previously found experimental results [157].

In another example, multi-metallic alloys have been investigated to predict the binding energy of CO [134]. Two sets of features have been developed in this study: 1) geometrical features such as local electronegativity and effective coordination number of adsorption sites, and 2) intrinsic properties of active sites on the catalysts surface such as ionic potential, electron affinity and Pauling electronegativity. Comparing the performance of the developed model in this study with models using d-band characteristics as features has shown that a high accuracy could be achieved just by using geometrical and intrinsic properties as features (Fig. 2.5). Although promising, this study is limited to a subset of multi-metallic alloys with only one surface termination.

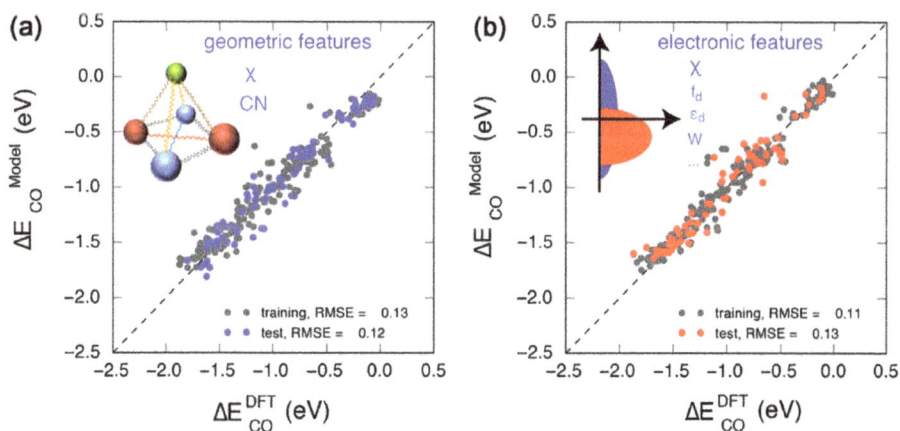

Fig. 2.5: Prediction accuracy using root mean square error (RMSE) as a metric for CO binding energies *based on the geometrical features local (atomic) electronegativity (χ) and effective coordination numbers (CN)* (a), and electronic structure features χ, d-band filling (f_d), ε_d, d-band width (W), d-band skewness, and d-band kurtosis for a database of multi-metallic alloys (b). Adapted from [100, 134] with permission from Elsevier.

An active learning model has also been used to screen a wide range of bimetallic alloys and identify optimal electrocatalysts for CO_2RR [151]. In this study, the authors have used a combination of a surrogate-based optimization that applies full-accuracy DFT calculations and a ML model to guide the DFT calculations. Different electronic and geometrical features for the active sites and neighboring elements coordinated with the adsorbates have been considered. These features include the atomic number of the elements coordinated with the adsorbates, the electronegativity of the element, the number of atoms of the element coordinated with the adsorbate, and the median adsorption energy between the adsorbate and the pure element. This model has been used to screen 17,507 unique surfaces with 1,684,908 unique adsorption sites and identify 131 candidate surfaces across 54 alloys for CO_2RR. After optimizing the surrogate models, a mean absolute error (MAE) value of 0.29 eV has been reported for the predicted CO binding energies with respect to the DFT-calculated binding energies. This method offers fully automated screening by taking advantage of ML and optimization to guide the DFT calculations and therefore reducing the number of required DFT calculations in the search for optimal catalysts. However, the heavy reliance on manual featurization used to guide and optimize active learning algorithms remains a challenge.

To enable the extraction of complex features from raw input instead of manually engineered features, several deep learning models that use multiple layers of neural networks have been developed [158]. These deep learning models use as input the atomic properties of the catalyst that do not require DFT calculations. Examples of previously developed deep learning models for catalysis include the crystal graph convolutional neural network (CGCNN), lattice convolutional neural network (LCNN) [158], atom centered symmetry function (ACSF) [118], and the deep learning architecture for molecules and materials SciNet [128].

The CGCNN takes advantage of a deep-learning convolutional neural network on top of a graph representation of materials. The graph representation consists of nodes that represent the constituent atoms of the material where different atomic attributes can be embedded and edges that represent the bonds formed between neighboring atoms (Fig. 2.6) [131]. In CGCNN, atoms can be connected with more than one bond indicating multiple edges among the nodes of the graph. The convolutional operation is performed several times on each bond that is a convolution of the atom features with its neighbors. The output of the convolutional operation is mapped out to the crystal properties. The CGCNN has shown great promise for predicting wide range of properties of bulk crystals, such as formation energy, Fermi energy, and bandgap. CGCNN only uses structural and basic atomic properties of the constituent elements, and none of the electronic structure features such as d-band characteristics are used. Efforts have been made to improve CGCNN by the either inclusion of new physical attributes or further optimization of the CGCNN [159]. For example, Karamad et al have recently developed orbital graph convolutional neural network (OGCNN) by inclusion of orbital–orbital interactions into the CGCNN [139]. It has been shown that the OGCNN outperforms the CGCNN by ~12–55%, when predicting the properties of different crystalline systems.

(a)

(b)

R Conv   L₁ hidden   Pooling   L₂ hidden

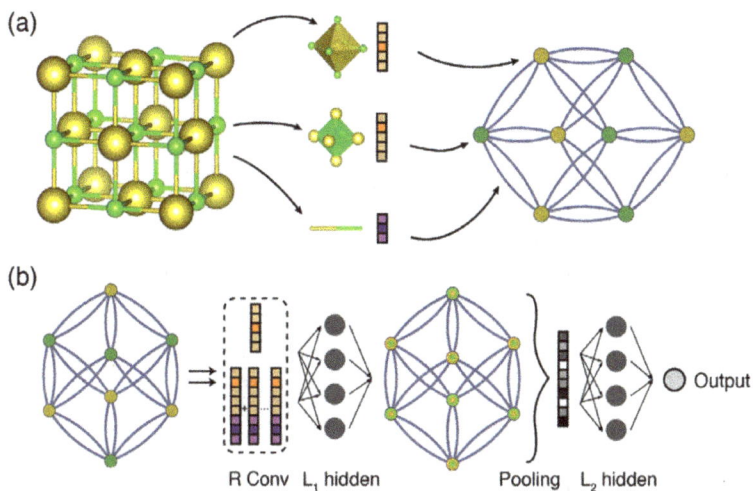

Fig. 2.6: (a) Nodes and edges represent atoms and bonds, respectively, when crystals are converted to graphs. (b) Architecture of a CGCNN. Adapted from [127] with permission from the American Physical Society.

Although CGCNN has originally been designed for predicting crystalline materials properties, it has been modified to be used for catalyst properties prediction [137, 140]. Back et al. have recently modified CGCNN to predict the binding energies of CO on diverse surfaces of intermetallic alloys. In the modified CGCNN, the role of adsorbate atoms on the surface structures are taken into account [137]. In addition, they have introduced "connectivity distance information" that addresses the movement of adsorbate atoms on the surface during structure relaxation. The adsorbate movement occurs when the adsorbate binding site changes during structure relaxation. In addition to the basic atomic properties being used as atomic features, Voronoi polyhedra between each atom and its neighbors have been considered in order to capture the geometrical information of the structures as well as the neighboring site information (Fig. 2.7) [137].

The CGCNN model modified by Back *et al.* has been further trained to predict the binding energies of CO adsorbate using three different sets of structures [137]: 1) initial structures where the adsorbate is arbitrarily placed on the surface, 2) final structures where DFT calculations are performed to relax the structure in which case the binding of the adsorbate to the site changes the structure, and 3) initial structures with connectivity distance information on adsorbates. Back et al. have reported a prediction performance for the binding energy of CO with MAE values of 0.19, 0.15, and 0.13 eV when initial structures, initial structures with connectivity distance information, and final structures were used, respectively. One notable effort in this study was the prediction of the binding energies using only initial structures as input with basic atomic features and neighboring site features that can be obtained at no additional computational cost. The authors suggested this reduced

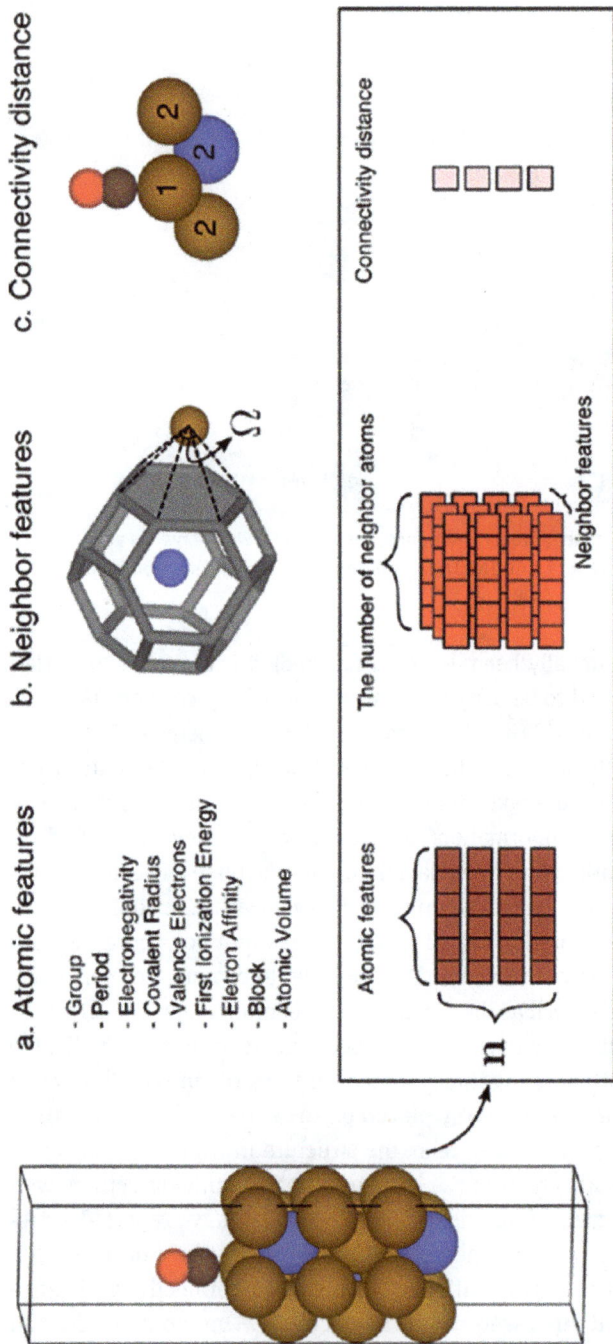

Fig. 2.7: Schematic of encoding a surface structure with CO as an adsorbate containing n atoms as input into the convolutional neural network: (a) Nine different basic atomic properties are used to represent each atom, similar to a CGCNN; (b) The geometrical information of the structures as well as the neighboring site information have been included by constructing Voronoi polyhedra between each atom and its neighbors; (c) Connectivity distances from adsorbate to all atoms in surface structures are counted. Adapted from [137] with permission from the American Physical Society.

computational cost approach to identify optimal catalysts via the high-throughput catalyst screening of promising candidate materials for full-accuracy DFT calculations [137].

Gu et al. have developed a novel approach called labeled site CGCNN (LS-CGCNN) for binding energy prediction where the adsorbate-interacting surface atoms in the initial unrelaxed structure are labeled [140]. This approach is realized with a lower computational cost because only the initial surface structures without adsorbates and DFT calculations are required to determine the surface structures with adsorbate. They have reported accuracy with MAE values of 0.128 eV for the binding energies of CO on diverse surfaces of intermetallic alloys. Further improvement of the LS-CGCNN approach has been achieved by the inclusion of an ensemble learning method for predicting the probability distribution of the binding energy. The latter has resulted in the prediction accuracy with a MAE value of 0.11 eV for the CO binding energy [140].

Gusarov et al. have proposed a reactivity descriptor for ML models based on the Fukui functions projected onto the Connolly surface and employed it to predict the adsorption energy of CO, a key mechanistic parameter of the CO_2RR on Ni and NiGa surfaces [160, 161]. The Connolly surface is formed by the smoothed van der Waals spheres of individual atoms. The projected Fukui functions capture the topology of a surface, an and important physicochemical property (equation (2.9)), providing a balance between accuracy and computational cost for screening the reactivity of a large number of materials [160].

The approach of projecting the 3D Fukui function on the 2D Connolly surface has important advantages over the projection of the Fukui function onto atomic sites (equations (2.15) and (2.16)) [150, 162]. The mapping to 2D accounts for the surface topology and different relative orientations of reactants near active sites [163]. The Fukui functions were calculated using DFT in periodic boundary conditions [160].

Figure 2.8 presents the vertical-colored contour plots of the Fukui function distribution in a plane perpendicular to the Ni (110) slab surface and passing through the axis of the CO molecule for several distances to the nearest Ni atom. The middle plots of Fig. 2.8 show that the redistribution of the Fukui function starts at distances approximately equivalent to the van der Waals radii [160]. This analysis agrees well with the study of electrophilic and nucleophilic attacks at the reaction center, featured by a small electrostatic force, on its van der Waals surface descriptors of nucleophilic and electrophilic regioselectivity [164].

Gusarov et al. have studied 12 surfaces of Ni and NiGa alloys with the most frequently used combinations of Miller indices and performed two types of calculations for each surface. First, they have performed a set of distinct geometry optimizations to map the binding energies of a CO molecule constrained along X and Y to a position shown in Fig. 2.9(a) and (b). The CO adsorption sites form a rectangular grid in the XY-plane (Fig. 2.9(a)). The adsorption energy contour plots are presented in Fig. 2.9(c) and (d) [160].

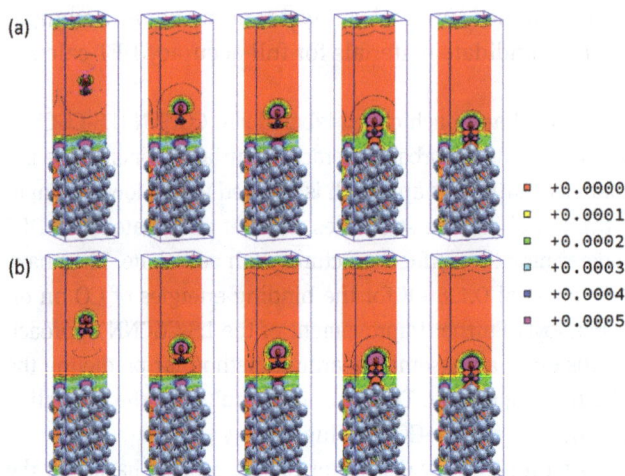

Fig. 2.8: The 2D redistribution of nucleophilic (a) and electrophilic (b) Fukui functions at distances between the CO molecule and the Ni (110) surface of 1.73, 2.23, 3.23, 3.73, and 7.0 Å. Adapted from [160] © 2020 Her Majesty the Queen in Right of Canada.

In the second type of calculation, they have computed the electrophilic and nucleophilic Fukui functions for bare surfaces using the finite-field method (equations (2.10) and (2.11) for derivatives from above and below taken at $\delta q = 0.5$) [53]. The calculated Fukui functions are approximated on the Connolly surface (Fig. 2.10) at the points corresponding to the tabulated map of CO binding energies.

Next, a linear regression analysis has been performed on the 12 Fukui functions projected onto the respective surfaces and the calculated 2D mapped binding energies [160]. While a strong correlation is obtained between projected Fukui functions and binding energies for an individual surface (Fig. 2.11(a)), a visible deviation is observed for a combined set of different slabs (Fig. 2.11(b)). The agreement is significantly improved by complementing the projected Fukui functions with the corresponding work functions (Fig. 2.11(c)) [160].

The approach of Gusarov et al. illustrates the effect of enhancing the local distribution by the global parameter, which defines the overall reaction direction [160]. A similar approach has been used by Gurkanet al. [165] and Damounet al. [166] to improve the performance of the Fukui functions in recognition of the reactivity of different sites in different molecules by multiplying the Fukui function by the global softness. However, in the study of Gusarov et al., the ML-based analysis benefited from the separate treatment of the Fukui function and the work function by using this specific algorithm [160]. The concept presented by Gusarov et al. could be improved further by convolving the 2D surface distribution of the Fukui function with specific local patterns representing the chemical structure of the reactants in order to enable the modeling of complex reactions involving the formation of several bonds [160].

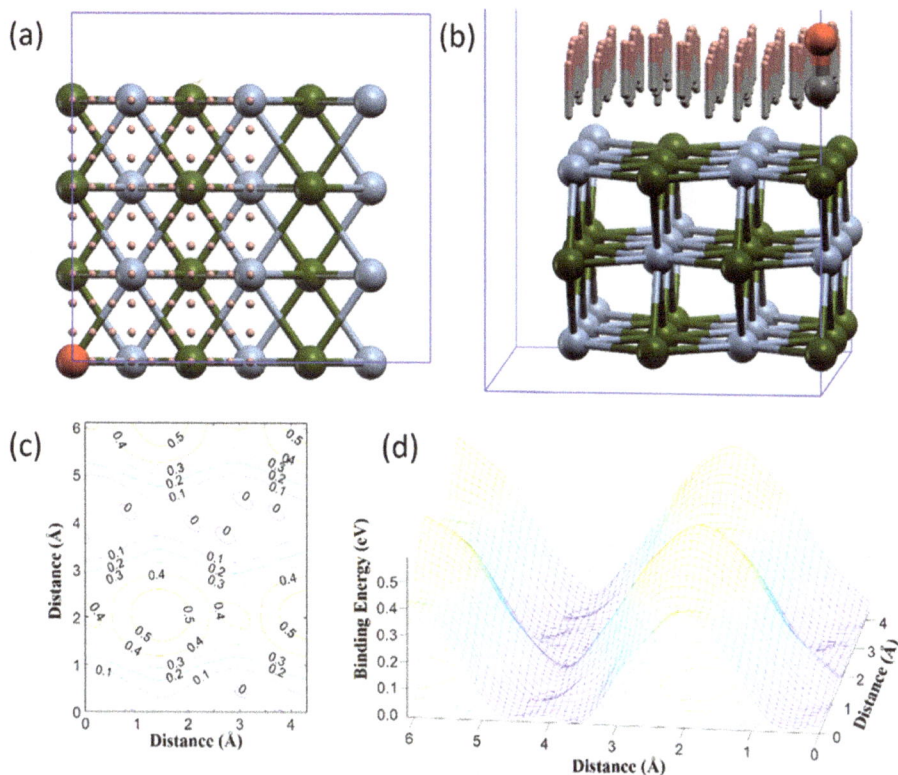

Fig. 2.9: Top (a) and side (b) views of NiGa (110) surface showing the positions of chemisorbed CO molecules; and contour plot (c) and 3D view (d) of the CO adsorption energies (in eV). Each CO position corresponds to a separate constrained optimization of a single CO molecule chemisorbed on the surface, as, for example, the molecule represented by ball and stick in the top right of (b). In the geometry optimizations, the X and Y coordinates of the C atom of CO are constrained. Adapted from [160] © 2020 Her Majesty the Queen in Right of Canada.

2.6.2 Oxygen evolution reaction

Water electrolysis is considered as a promising sustainable technology to produce hydrogen thereby reducing dependence on fossil fuels. The OER is the half reaction of water electrolysis. High performance inexpensive and efficient electrocatalysts materials are needed for the development of a feasible and sustainable water electrolysis technology. A combined approach of using ML models and DFT calculations can help to boost the development of novel catalysts.

Back et al. have recently developed an automated and high-throughput screening approach by combining DFT calculations and ML models to predict the electrocatalytic activity of metal oxides [138]. They have also identified the most stable

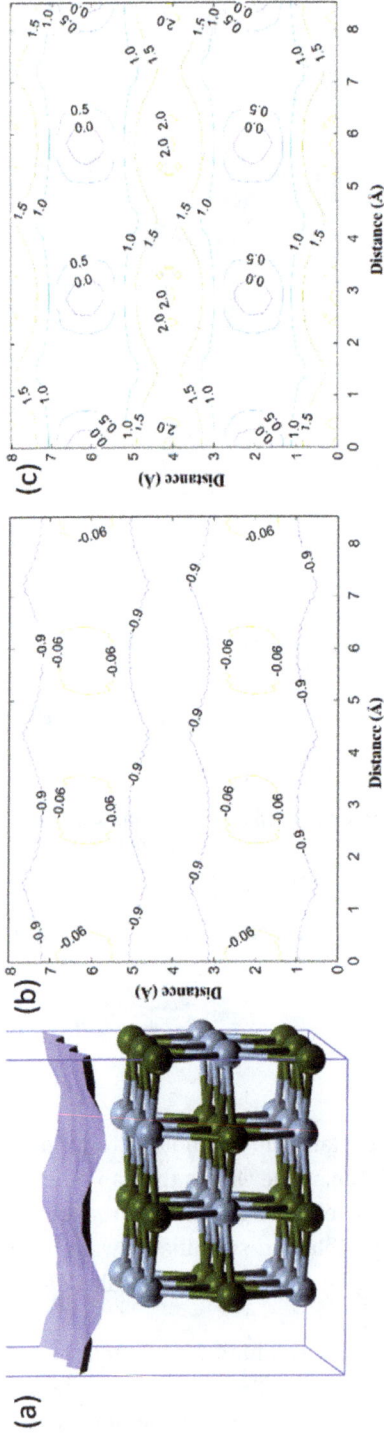

Fig. 2.10: The Connolly surface for NiGa (110) (a) and contour plots of projected nucleophilic (b) and electrophilic (c) Fukui functions. Adapted from [160] © 2020 Her Majesty the Queen in Right of Canada.

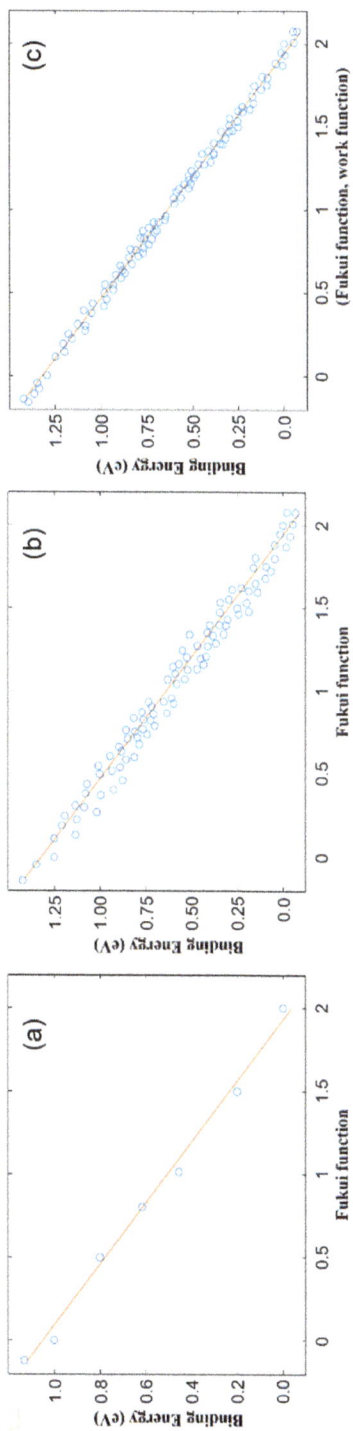

Fig. 2.11: Linear regression of projected Fukui function and CO adsorption energies onto Ni (110) surface (a); for the set of 12 studied surfaces, R = −0.92 (b); and for multivariable linear regression, where the Fukui function is augmented by the work function, R = −0.97 (c). Adapted from [160] © 2020 Her Majesty the Queen in Right of Canada.

surface coverage under relevant electrochemical conditions for OER. Different crystal structures and different surface terminations (Fig. 2.12(a)) have been considered to explore catalytic activity of various active sites on metal oxides. The CGCNN approach has been used to predict the binding energies based only on the initial unrelaxed structures. The reported MAE values are 0.07 eV for coverage calculations of O* and OH* adsorbates (Fig. 2.12(b)). Moreover, an MAE value of 0.13 eV for OER calculations is obtained using the predicted binding energies of O*, OH*, and OOH* (Fig. 2.12(c)) [138].

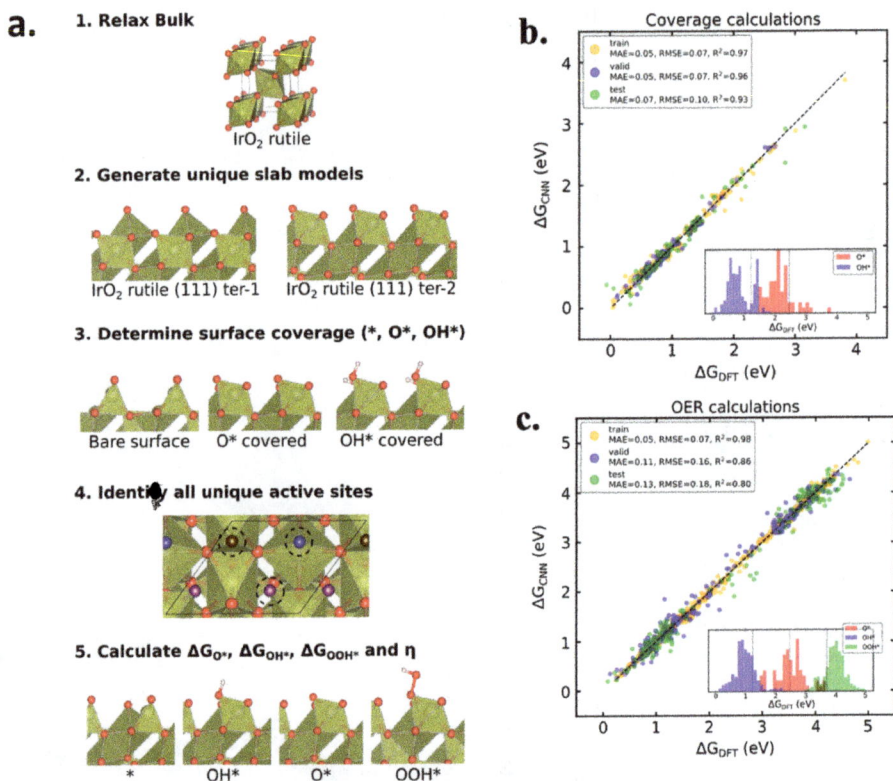

Fig. 2.12: (a) *Activity investigation steps for the OER catalyst* IrO$_2$ generating different surface terminations (1) from bulk (0), determining surface coverages of relevant adsorbates (3), identifying all unique active sites (4), and calculating the binding energies of involved intermediates in OER reaction and theoretical overpotentials (5). (b) Comparison of CGCNN-predicted coverage calculations for O* and OH* with DFT-calculated values. (c) Comparison of CGCNN-predicted OER calculations of O*, OH* and OOH* adsorbates with DFT-calculated values. Adapted from [138] with permission from the American Chemical Society.

2.6.3 Methanol electrooxidation

The direct methanol fuel cell is an attractive clean power source for stationary and portable devices due to its high efficiency and low emission of pollutants [167]. Catalyst development is at the heart of the methanol electrooxidation technology. In recent years, different ML models combined with DFT calculated linear scaling relations have been used to screen a library of bimetallic alloys in order to identify optimal catalysts for methanol electrooxidation [168]. In the work of Li *et al.* [168], ML models have been trained to predict the binding energies of CO and OH as the reactivity descriptors that govern the methanol electrooxidation catalytic activity. Both primary and secondary features have been used to train these ML models. The electronic structure-based features, including *d*-band characteristics and geometry-based features (calculated as the geometric means of the electronegativity of metals atoms within certain radius) have been used as the primary features. The basic atomic properties of active sites on the catalytic surface, such as ionic potential, work function, and atomic radius, have been employed as the secondary features. Different ML models including linear regression, least absolute shrinkage and selection operator (LASSO), neural networks, and random forest have been utilized to train the input data. Using this approach, several known and new catalysts for methanol electrooxidation have been identified.

2.7 Conclusions

The ML-based methods employed in AI offer a promising approach to address problems in heterogeneous catalysis. There are, however, few shortcomings in the realm of ML for catalysis that need to be addressed. Firstly, the development of new databases that span a wider chemical space than existing one is critical. The existing catalysis databases are limited in terms of the classes of materials preventing the deployment of ML models to search among new classes of catalysts. Secondly, an appropriate combination of global and local parameters is critical for the construction of accurate and meaningful ML descriptors in order to map the most important features of heterogeneous catalysis. With the correct choice of descriptors, global parameters define a general direction for a reaction, whereas the local parameters specify the particular interaction and follow the redistribution of electronic density, helping to compare the different sites of the heterogeneous molecular system [165]. For example, modern ML algorithms allow us to take non-linear associations and higher-order interactions into account automatically and to resolve the complex relations between Fukui functions and reactivity. Thirdly, although significant progress has been achieved in the development of ML models with special abilities to extract features using only raw materials data and with high predictive accuracy, these ML models often lack interpretability. Therefore, it is necessary to develop more sophisticated ML models with enhanced interpretability that can provide additional physical and chemical information for accelerated catalysis discovery.

References

[1]   Wisniak J. The history of catalysis. From the beginning to Nobel Prizes. Educ Quím, 2010, 21, 60–69.

[2]   Rothenberg G. Catalysis: Concepts and Green Applications. Wiley-VCH, 2008, 65.

[3]   Franceschetti AG, Zunger A. The inverse band-structure problem of finding an atomic configuration with given electronic properties. Nature, 1999, 402, 60–63.

[4]   Ceder G, Chiang Y-M, Sadoway DR, Aydinol MK, Jang YI, Huang B. Identification of cathode materials for lithium batteries guided by first-principles calculations. Nature, 1998, 392, 694–696.

[5]   Jóhannesson GH, Bligaard T, Ruban AV, Skriver HL, Jacobsen KW, Nørskov JK. Combined electronic structure and evolutionary search approach to materials design. Phys Rev Lett, 2002, 88, 255506.

[6]   Castelli IE, Landis DD, Thygesen KS, Dahl S, Chorkendorff I, Jaramillo TF, Jacobsen KW. New cubic perovskites for one- and two-photonwater splitting using the computational materials repository. Energy Environ Sci, 2012, 5, 9034–9043.

[7]   Hohenberg P, Kohn W. Inhomogeneous electron gas. Phys Rev, 1964, 136, B864–B871.

[8]   Mamun O, Winther KT, Boes JR, Bligaard T. High-throughput calculations of catalytic properties of bimetallic alloy surfaces. Sci Data, 2019, 6, 76.

[9]   Perez AE, Ribadeneira R. Modeling with DFT and Chemical Descriptors Approach for the Development of Catalytic Alloys for PEMFCs. In: Density Functional Theory (Chapter 2), IntechOpen, 2019.

[10]   Nørskov JK, Pedersen FA, Studt F, Bligaard T. Density functional theory in surface chemistry and catalysis. Proc Natl Acad Sci USA, 2011, 108, 937–943.

[11]   Ghiringhelli LM, Vybiral J, Levchenko SV, Draxl C, Scheffler M. Big data of materials science: critical role of the descriptor. Phys Rev Lett, 2015, 114, 105503.

[12]   Seko A, Togo A, Hayashi H, Tsuda K, Chaput L, Tanaka I. Prediction of low-thermal-conductivity compounds with first-principles anharmonic lattice-dynamics calculations and Bayesian optimization. Phys Rev Lett, 2015, 115, 205901.

[13]   Deml AM, O'Hayre R, Wolverton C, Stevanovic V. Predicting density functional theory total energies and enthalpies of formation of metal-nonmetal compounds by linear regression. Phys Rev B, 2016, 93, 085142.

[14]   Meredig B, Agrawal A, Kirklin S, Saal JE, Doak JW, Thompson A, Zhang K, Choudhary A, Wolverton C. Combinatorial screening for new materials in unconstrained composition space with machine learning. Phys Rev B, 2014, 89, 094104.

[15]   Dey P, Bible J, Datta S, Broderick S, Jasinski J, Sunkara M, Menon M, Rajan K. Informatics-aided bandgap engineering for solar materials. Comput Mater Sci, 2014, 83, 185–195.

[16]   Xue D, Balachandran PV, Hogden J, Theiler J, Xue D, Lookman T. Accelerated search for materials with targeted properties by adaptive design. Nature Commun, 2016, 7, 11241.

[17]   Isayev O, Oses C, Toher C, Gossett E, Curtarolo S, Tropsha A. Universal fragment descriptors for predicting properties of inorganic crystals. Nat Commun, 2017, 8, 15679.

[18]   Zhou Q, Tang P, Liu S, Pan J, Yan Q, Zhang S-C. Learning atoms for materials discovery. Proc Nat Acad Sci USA, 2018, 115, E6411–7.

[19]   Ma X, Li Z, Achenie LEK, Xin H. Machine-learning-augmented chemisorption Model for CO2 electroreduction catalyst screening. J Phys Chem Lett, 2015, 6, 3528–3533.

[20]   Schmidt J, Shi J, Borlido P, Chen L, Botti S, Marques MAL. Predicting the thermodynamic Stability of solids combining density functional theory and machine learning. Chem Mater, 2017, 29, 5090–5103.

[21]   Liu Y, Zhao T, Ju W, Shi S. Materials discovery and design using machine learning. J Materiomics, 2017, 3, 159–177.

[22]   Hattori T, Kito S. Neural network as a tool for catalyst development. Catal Today, 1995, 23, 347–355.

[23]   Sasaki M, Hamada H, Kintaichi Y, Ito T. Application of a neural network to the analysis of catalytic reactions analysis of NO decomposition over Cu/ZSM-5 zeolite. Appl Catal, A, 1995, 132, 261–270.

[24]   Lamoureux PS, Winther KT, Torres JAG, Streibel V, Zhao M, Bajdich M, Abild-Pedersen F, Bligaard T. Machine learning for computational heterogeneous catalysis. ChemCatChem, 2019, 11, 3581–3601.

[25]   McCullough K, Williams T, Mingle K, Jamshidi P, Lauterbach J. High-throughput experimentation meets artificial intelligence: A new pathway to catalyst discovery. Phys Chem Chem Phys, 2020, 22, 11174–11196.

[26]   Gusarov S, Dmitriev YY, Stoyanov SR, Kovalenko A. Koopmans' multiconfigurational self-consistent field (MCSCF) Fukui functions and MCSCF perturbation theory. Can J Chem, 2013, 91, 886–893.

[27]   Van Santen RA, Neurock M. Concepts in theoretical heterogeneous catalytic reactivity. Catal Rev Sci Eng, 1995, 37, 557–698.

[28]   Harvey JN, Himo F, Maseras F, Perrin L. Scope and challenge of computational methods for studying mechanism and reactivity in homogeneous catalysis. ACS Catal, 2019, 9, 6803–6813.

[29]   Tsang AS-K, Sanhueza IA, Schoenebeck F. Combining experimental and computational studies to understand and Predict reactivities of relevance to homogeneous catalysis. Chem Eur J, 2014, 20, 16432–16441.

[30]   Nascimento MAC, ed. Theoretical Aspects of Heterogeneous Catalysis, Dordrecht, Netherlands, Kluwer Acad. Publ., 2001.

[31]   Falivene L, Kozlov SM, Cavallo L. Constructing bridges between computational tools in heterogeneous and homogeneous catalysis. ACS Catal, 2018, 8, 6, 5637–5656.

[32]   Kumar KV, Gadipelli S, Wood B, Ramisetty KA, Stewart AA, Howard CA, Brett DJL, Rodriguez-Reinoso F. Characterization of the adsorption site energies and heterogeneous surfaces of porous materials. J Mater Chem A, 2019, 7, 10104–10137.

[33]   Zhu S, Huang X, Ma F, Wang L, Duan X, Wang S. Catalytic removal of aqueous contaminants on N-Doped graphitic biochars: inherent roles of adsorption and nonradical mechanisms. Environ Sci Technol, 2018, 52, 15, 8649–8658.

[34]   Kasemo B. Biological surface science. Surf Sci, 2002, 500, 656–677.

[35]   Yokoyama T, Yokoyama S, Kamikado T, Okuno Y, Mashiko S. Selective assembly on a surface of supramolecular aggregates with controlled size and shape. Nature, 2001, 413, 619–621.

[36]   Gray MR, Tykwinski RR, Stryker JM, Tan X. Supramolecular assembly model for aggregation of petroleum asphaltenes. Energy Fuels, 2011, 25, 7, 3125–3134.

[37]   Whitesides GM, Boncheva M. Beyond molecules: self-assembly of mesoscopic and macroscopic components. Proc Natl Acad Sci USA, 2002, 99, 4769–4774.

[38]   Pearson RG. Hard and soft acids and bases. J Am Chem Soc, 1963, 85, 3533–3539.

[39]   Pearson RG. Acids and bases. Science, 1966, 151, 172–177.

[40]   Lewis GN. Valence and the Structure of Atoms and Molecules, Chemical Catalog Company, New York, NY, USA, 1923, 142.

[41]   Parr RG, Pearson RG. Absolute hardness: companion parameter to absolute electronegativity. J Am Chem Soc, 1983, 105, 7512–7516.

[42]   Parr RG, Donnely RA, Levy M, Palke WE. Electronegativity: the density functional viewpoint. J Chem Phys, 1978, 68, 3801.

[43]   Parr RG, Yang WJ. Density functional approach to the frontier-electron theory of chemical reactivity. J Am Chem Soc, 1984, 106, 4049–4050.

[44]   Sanderson RT. An interpretation of bond lengths and a classification of bonds. Science, 1951, 114, 670–672.

[45]   Pearson RG. The principle of maximum hardness. Acc Chem Res, 1993, 26, 5, 250–255.

[46]   Parr RG, Chattaraj PK. Principle of maximum hardness. J Am Chem Soc, 1991, 113, 5, 1854–1855.

[47]   Yang W, Zhang Y, Ayers PW. Degenerate ground states and a fractional number of electrons in density and reduced density matrix functional theory. Phys Rev Lett, 2000, 84, 5172–5175.

[48]   Parr RG, Bartolotti LJ. On the geometric mean principle for electronegativity equalization. J Am Chem Soc, 1982, 104, 3801–3803.

[49]   Ayers PW, Parr RG, Pearson RG. Elucidating the hard/soft acid/base principle: A perspective based on half-reactions. J Chem Phys, 2006, 124, 194107.

[50]   Parr RG, Yang W. Density-functional theory of the electronic structure of molecules. Annu Rev Phys Chem, 1995, 46, 701–728.

[51]   Fukui K, Yonezawa T, Shingu HA. Molecular orbital theory of reactivity in aromatic hydrocarbons. J Chem Phys, 1952, 20, 722–725.

[52]   Fukui K, Yonezawa T, Nagata C, Shingu H. Molecular orbital theory of orientation in aromatic, heteroaromatic, and other conjugated molecules. J Chem Phys, 1954, 22, 1433–1442.

[53]   DMol³. Materials Studio 5.0, San Diego, CA, USA, Accelrys Inc., 2009.

[54]   Perdew JP, Parr RG, Levy M, Balduz JL Jr. Density-functional theory for fractional particle number: derivative discontinuities of the energy. Phys Rev Lett, 1982, 49, 1691–1694.

[55]   Fukui K, Yonezawa T, Nagata C. Theory of substitution in conjugated molecules. Bull Chem Soc Jap, 1954, 27, 423–427.

[56]   Fukui K. Role of frontier orbitals in chemical reactions. Science, 1987, 218, 747–754.

[57]   Bartolotti LJ, Ayers PW. An example where orbital relaxation is an important contribution to the fukui function. J Phys Chem A, 2005, 109, 1146–1151.

[58]   Ayers PW, Levy M. Perspective on "Density Functional Approach to the Frontier-Electron Theory of Chemical Reactivity.". Theor Chem Acc, 2000, 103, 353–360.

[59]   Sablon N, De Proft F, Ayers PW. Computing fukui functions without differentiating with respect to electron number. II. Calculation of condensed molecular fukui functions. J Chem Phys, 2007, 126, 224108.

[60]   Stoyanov SR, Yin C-X, Gray MR, Stryker JM, Gusarov S, Kovalenko A. Computational and experimental study of the structure, binding preferences, and spectroscopy of Nickel(II) and Vanadyl porphyrins in petroleum. J Phys Chem B, 2010, 114, 2180–2188.

[61]   Treibs A. Über das Vorkommen von Chlorophyllderivaten in einem Ölschiefer aus der oberen trias *(Occurrence of Chlorophyll Derivatives in an Oil Shale of the Upper Triassic)*. Liebigs Ann, 1934, 509, 103–114.

[62]   Chauhan G, De Klerk A. Acidified ionic liquid assisted recovery of vanadium and nickel from oilsands bitumen. Energy Fuels, 2021, 35, 5963–5974.

[63]   Stoyanov SR, Yin C-X, Gray MR, Stryker JM, Gusarov S, Kovalenko A. Density functional theory investigation of the effect of axial coordination and annelation on the absorption

spectroscopy of nickel(II) and vanadyl porphyrins relevant to bitumen and crude oils. Can J Chem 2013, 91, 872–8.

[64]   Rapier R. Why vanadium flow batteries may be the future of utility-scale energy storage. Forbes.com, 2020, (Accessed March 31, 2021, at https://www.forbes.com/sites/rrapier/2020/10/24/why-vanadium-flow-batteries-may-be-the-future-of-utility-scale-energy-storage/?sh=2a83c5a82305.

[65]   Stoyanov SR, Gusarov S, Kovalenko A. Modelling of bitumen fragment adsorption on Cu+ and Ag+ exchanged zeolite nanoparticles. Molec Simul, 2008, 34, 943–951.

[66]   Klopman G. Chemical reactivity and the concept of charge- and frontier-controlled reactions. J Am Chem Soc, 1968, 90, 223–234.

[67]   Cardenas C, Tiznado W, Ayers PW, Fuentealba P. The fukui potential and the capacity of charge and the global hardness of atoms. J Phys Chem A, 2011, 115, 2325–2331.

[68]   Cardenas C, Rabi N, Ayers PW, Morell C, Jaramillo P, Fuentealba P. Chemical reactivity descriptors for ambiphilic reagents: dual descriptor, local hypersoftness, and electrostatic potential. J Phys Chem A, 2009, 113, 8660–8667.

[69]   Morell C, Hocquet A, Grand A, Torro-Labbe A. New dual descriptor for chemical reactivity. J Phys Chem A, 2005, 109, 205–212.

[70]   Cho M, Sylvetsky N, Eshafi S, Santra G, Efremenko I, Martin JML. The atomic partial charges arboretum: trying to see the forest for the trees. ChemPhysChem, 2020, 21, 688–696.

[71]   Guerra SF, Handgraaf J-W, Baerends EJ, Bickelhaupt FM. Voronoi deformation density (VDD) charges: assessment of the Mulliken, Bader, Hirshfeld, Weinhold, and VDD methods for charge analysis. J Comput Chem, 2004, 25, 189–210.

[72]   Parr RG, Yang W. Density-Functional Theory of Atoms and Molecules. New York, NY, USA, Oxford University Press, 1989.

[73]   Elanany M, Su B-L, Vercauteren DP. The effect of framework organic moieties on the acidity of zeolites: A DFT study. J Mol Catal A: Chem, 2007, 263, 195–199.

[74]   Chatterjee A. A reactivity index study to rationalize the effect of dopants on Brönsted and Lewis acidity occurring in MeAlPOs. J Mol Graphics Model, 2006, 24, 262–270.

[75]   RCh D, Roy RK, Hirao K. Local reactivity descriptors to predict the strength of Lewis acid sites in alkali cation-exchanged zeolites. Chem Phys Lett, 2004, 389, 186–190.

[76]   Stoyanov SR, Gusarov S, Kuznicki SM, Kovalenko A. Theoretical modeling of zeolite nanoparticle surface acidity for heavy oil upgrading. J Phys Chem C, 2008, 112, 6794–6810.

[77]   Roy RK, Hirao K, Krishnamurthy S, Pal S. Mulliken population analysis based evaluation of condensed Fukui function indices using fractional molecular charge. J Chem Phys, 2001, 115, 2901–2907.

[78]   Bultinck P, Van Alsenoy C, Ayers PW, Carbó-Dorca R. Critical analysis and extension of the hirshfeld atoms in molecules. J Chem Phys, 2007, 126, 144111.

[79]   Yang WJ, Parr RG. Hardness, softness, and the fukui function in the electronic theory of metals andcatalysis. Proc Natl Acad Sci USA, 1985, 82, 6723–6726.

[80]   Allison TC, Tong YJ. Application of the condensed fukui function to predict reactivity in core–shell transition metal nanoparticles. Electrochim Acta, 2012, 101, 334–340.

[81]   Pearson RG. Hard and Soft Acids and Bases. Dowden, Hutchinson and Ross, Stroudsburg, PA, USA, 1973, 3536.

[82]   Ayers PW, Melin J. Computing the fukui function from ab initio quantum chemistry: approaches based on the extended Koopmans' theorem. Theor Chem Acc, 2007, 117, 371–381.

[83]   Gusarov S, Fedorova T, Dmitriev Y, Kovalenko A. On variational estimates for exchange-correlation interaction obtained within super-CI approach to MCSCF approximation. Int J Quant Chem, 2009, 109, 1672–1675.

[84] Gusarov S, Goidenko I, Dmitriev Y, Labzowsky L. Variational estimates for exchange-correlation interaction obtained within super-CI approach to MCSCF approximation. Int J Quant Chem, 2007, 107, 2616–2621.

[85] Gu Y, Xu X. Extended Koopmans' theorem in the adiabatic connection formalism: applied to doubly hybrid density functionals. J Chem Phys, 2020, 153, 044109.

[86] Gu Y, Xu X. Extended Koopmans' theorem at the second-order perturbation theory. J Comput Chem, 2020, 41, 1165–1174.

[87] Pino-Rios R, Inostroza D, Cárdenas-Jirón G, Tiznado W. Orbital-weighted dual descriptor for the study of local reactivity of systems with (Quasi-) degenerate states. J Phys Chem A, 2019, 123, 49, 10556–10562.

[88] Bultinck P, Clarisse D, Ayers PW, The C-DR. Fukui matrix: A simple approach to the analysis of the fukui function and its positive character. Phys Chem Chem Phys, 2011, 13, 6110–6115.

[89] Bultinck P, Cardenas C, Fuentealba P, Johnson PA, Ayers PW. How to compute the fukui matrix and function for systems with (Quasi-)degenerate states. J Chem Theory Comput, 2014, 10, 202–210.

[90] Morgenstern A, Wilson TR, Eberhart ME. Predicting chemical reactivity from the charge density through gradient bundle analysis: moving beyond fukui functions. J Phys Chem A, 2017, 121, 22, 4341–4351.

[91] Stuyver T, De Proft F, Geerlings P, Shaik S. How do local reactivity descriptors shape the potential energy surface associated with chemical reactions? The valence bond delocalization perspective. J Am Chem Soc, 2020, 142, 22, 10102–10113.

[92] Hammer B, Nørskov JK. Electronic factors determining the reactivity of metal surfaces. Surf Sci, 1995, 343, 211–220.

[93] Hammer B, Nørskov JK. Why gold is the noblest of all the metals. Nature, 1995, 376, 238–240.

[94] Hammer B, Nørskov JK. Theoretical surface science and catalysis – calculations and concepts. Adv Catal, 2000, 45, 71–129.

[95] Hammer B, Nørskov JK. Why gold is the noblest of all the metals. Nature, 1995, 376, 238–240.

[96] Greeley J, Nørskov JK, Mavrikakis M. Electronic structure and catalysis on metal surfaces. Annu Rev Phys Chem, 2002, 53, 319–348.

[97] Hammer B, Nørskov JK. Electronic factors determining the reactivity of metal surfaces. Surf Sci, 1995, 343, 211–220.

[98] Hammer B, Nørskov JK. Theoretical surface science and catalysis – calculations and concepts Adv Catal, 2000, 45, 71–129.

[99] Bhattacharjee S, Waghmare U, Lee S. An improved d-band model of the catalytic activity of magnetic transition metal surfaces. Sci Rep, 2016, 6, 35916.

[100] Ulissi ZW, Tang MT, Xiao J, Liu X, Torelli DA, Karamad M, Cummins K, Hahn C, Lewis NS, Jaramillo TF, Chan K, Nørskov JK. Machine-learning methods enable exhaustive searches for active bimetallic facets and reveal active site motifs for CO2 reduction. ACS Catal, 2017, 7, 6600–6608.

[101] Goldsmith BR, Esterhuizen J, Liu J-X, Bartel CJ, Sutton C. Machine learning for heterogeneous catalyst design and discovery. AIChE J, 2018, 64, 2311–2323.

[102] Mueller T, Kusne AG, Ramprasad R. Machine Learning in Materials Science: Recent Progress and Emerging Applications. In Parrill AL, Lipkowitz KB, eds. Reviews in Computational Chemistry 29, Wiley, 2016, 186–273.

[103] Miura K, Watanabe R, Fukuhara C. Theoretical study of oxygen adsorption energy on supported metal clusterusing d-band center theory and HSAB concept. Surf Sci, 2020, 696, 121601.

[104] Chen J, Meng D, Du J. Stability and local reactivity of Pd-Au nanoclusters. Molec Phys, 2017, 115, 1475–1479.

[105] Gao W, Chen Y, Li B, Liu S-P, Liu X, Jiang Q. Determining the adsorption energies of small molecules with the intrinsic properties of adsorbates and substrates. Nature Commun, 2020, 11, 1196.

[106] Zakutayev A, Wunder N, Schwarting M, Perkins JD, White R, Munch K, Tumas W, Phillips C. An open experimental database for exploring inorganic materials. Sci Data, 2018, 5, 180053.

[107] Rothenberg G. Data mining in catalysis: separating knowledge from garbage. Catal Today, 2008, 137, 2–10.

[108] Jain A, Ong SP, Hautier G, Chen W, Richards WD, Dacek S, Cholia S, Gunter D, Skinner D, Ceder G, Persson KA. Commentary: the materials project: A materials genome approach to accelerating materials innovation. APL Mater, 2013, 1, 011002.

[109] Belsky A, Hellenbrandt M, Karen VL, Luksch P. New developments in the inorganic crystal structure database (ICSD): accessibility in support of materials research and design. Acta Crystallogr Sect B, 2002, 58, 364–369.

[110] Draxl C, Scheffler M. NOMAD: The FAIR concept for big data-driven materials science. MRS Bull, 2018, 43, 676–682.

[111] Curtarolo S, Setyawan W, Wang S, Xue J, Yang K, Taylor RH, Nelson LJ, Hart GL, Sanvito S, Buongiorno-Nardelli M, Mingo N, Levyh O. AFLOWLIB.ORG: A distributed materials properties repository from high-throughput ab initio calculations. Comput Mater Sci, 2012, 58, 227–235.

[112] Saal JE, Kirklin S, Aykol M, Meredig B, Wolverton C. Materials design and discovery with high-throughput density functional theory: the open quantum materials database (OQMD). JOM, 2013, 65, 1501–1509.

[113] Landis DD, Hummelshoj JS, Nestorov S, Greeley J, Dulak M, Bligaard T, Norskov JK, Jacobsen KW. Comput Sci Eng, 2012, 14, 51–57.

[114] Chanussot L, Das A, Goyal S, Lavril T, Shuaibi M, Riviere M, Tran K, Heras-Domingo J, Ho C, Hu W, Palizhati A, Sriram A, Wood B, Yoon J, Parikh D, Zitnick CL, Ulissi Z. The open catalyst 2020 (OC20) dataset and community challenges. arXiv:2010.09990v4 (cond-mat.mtrl-sci) Mar 16, 2021, ver. 4.

[115] Winther K, Hoffmann MJ, Mamun O, Boes JR, Nørskov JK, Bajdich M, Bligaard T. Catalysis-Hub.org, an open electronic structure database for surface reactions. Sci Data, 2019, 6, 75.

[116] Hummelshøj JS, Abild-Pedersen F, Studt F, Bligaard T, Nørskov JK. CatApp: A web application for surface chemistry and heterogeneous catalysis. Angew Chem Int Ed, 2012, 51, 272–274.

[117] Behler J, Parrinello M. Generalized neural-network representation of high-dimensional potential-energy surfaces. Phys Rev Lett, 2007, 98, 146401.

[118] Behler J. Atom-centered symmetry functions for constructing high-dimensional neural network potentials. J Chem Phys, 2011, 134, 074106.

[119] Bartók AP, Kondor R, Csányi G. On representing chemical environments. Phys Rev B, 2013, 87, 184115.

[120] Huo H, Rupp M. Unified representation of molecules and crystals for machine learning, arXiv:1704.06439v3 (physics.chem-ph) Jan 2, 2018, ver. 3.

[121] Rupp M, Tkatchenko A, Müller KR, Von Lilienfeld OA. Fast and accurate modeling of molecular atomization energies with machine learning. Phys Rev Lett, 2012, 108, 058301.

[122] Isayev O, Fourches D, Muratov EN, Oses C, Rasch K, Tropsha A, Curtarolo S. Materials cartography: representing and mining materials space using structural and electronic fingerprints. Chem Mater, 2015, 27, 735–743.

[123] Ward L, Agrawal A, Choudhary A, Wolverton C. A general-purpose machine learning framework for predicting properties of inorganic materials. NPJ Comput Mater, 2016, 2, 16028.

[124] Ye W, Chen C, Wang Z, Chu IH, Ong SP. Deep neural networks for accurate predictions of crystal stability. Nat Commun, 2018, 9, 3800.

[125] Schütt KT, Glawe H, Brockherde F, Sanna A, Müller KR, Gross EKU. Structural representations, on the other hand, encode local chemical environments of atoms by capturing the geometry and interaction between atoms. Phys Rev B, 2014, 89, 205118.

[126] Pham TL, Kino H, Terakura K, Miyake T, Tsuda K, Takigawa I, Dam HC. Machine learning reveals orbital interaction in materials. Sci Technol Adv Mater, 2017, 18, 756–765.

[127] Xie T, Grossman JC. Crystal graph convolutional neural networks for an accurate and interpretable prediction of material properties. Phys Rev Lett, 120, 145301.

[128] Schütt KT, Sauceda HE, Kindermans P-J, Tkatchenko A, Müller K-R. SchNet – A deep learning architecture for molecules and materials. J Chem Phys, 2018, 148, 241722.

[129] Noh J, Kim J, Back S, Jung Y. Catalyst design using actively learned machine with non-ab initio input features towards CO2 reduction reactions. arXiv:1709.04576v1 (cond-mat.mtrl-sci) Sep 14, 2017, ver. 1.

[130] Takigawa I, Shimizu K, Tsuda K, Takakusagi S. Machine-learning prediction of the d-band center for metals and bimetals. RSC Adv, 2016, 6, 52587–52595.

[131] Hong WT, Risch M, Stoerzinger KA, Grimaud A, Suntivich J, Shao-Horn Y. Toward the rational design of non-precious transition metal oxides for oxygen electrocatalysis. Energy Environ Sci, 2015, 8, 1404–1427.

[132] Dickens CF, Montoy JH, Kulkarni AR, Bajdich M, Nørskov JK. An electronic structure descriptor for oxygen reactivity at metal and metal-oxide surfaces. Surf Sci, 2019, 681, 122–129.

[133] Choksi TS, Roling LT, Streibel V, Abild-Pedersen F. Predicting adsorption properties of catalytic descriptors on bimetallic nanoalloys with site-specific precision. J Phys Chem Lett, 2019, 10, 1852–1859.

[134] Li Z, Ma X, Xin H. Feature engineering of machine-learning chemisorption models for catalyst design. Catal Today, 2017, 280, 232–238.

[135] Roling LT, Abild-Pedersen F. Structure-sensitive scaling relations: adsorption energies from surface site stability. ChemCatChem, 2018, 10, 1643–1650.

[136] Krizhevsky A, Sutskever I, Hinton GE. ImageNet classification with deep convolutional neural networks. Commun Assoc Comput Machin, 2017, 60, 84–90.

[137] Back S, Yoon J, Tian N, Zhong W, Tran K, Ulissi ZW. Convolutional neural network of atomic surface structures to predict binding energies for high-throughput screening of catalysts. J Phys Chem Lett, 2019, 10, 4401–4408.

[138] Back S, Tran K, Ulissi ZW. Toward a design of active oxygen evolution catalysts: insights from automated density functional theory calculations and machine learning. ACS Catal, 2019, 9, 7651–7659.

[139] Karamad M, Magar R, Shi Y, Siahrostami S, Gates ID, Farimani AB. Orbital graph convolutional neural network for material property prediction. Phys Rev Mater, 2020, 4, 093801.

[140] Gu GH, Noh J, Kim S, Back S, Ulissi Z, Jung Y. Practical deep-learning representation for fast heterogeneous catalyst screening. J Phys Chem Lett, 2020, 11, 3185–3191.

[141] Wu Z, Ramsundar B, Feinberg EN, Gomes J, Geniesse C, Pappu AS, Leswing K, Pande V. MoleculeNet: A benchmark for molecular machine learning. Chem Sci, 2018, 9, 513–530.

[142] Gu GH, Choi C, Lee Y, Situmorang AB, Noh J, Kim YH, Jung Y. Progress in computational and machine-learning methods for heterogeneous small-molecule activation. Adv Mater, 2020, 32, 1907865.

[143] Tran K, Neiswanger W, Yoon J, Xing E, Ulissi ZW. Methods for comparing uncertainty quantifications for material property predictions. Machine Learning: Sci Technol, 2020, 1, 025006.

[144] Toyao T, Suzuki K, Kikuchi S, Takakusagi S, Shimizu K, Takigawa I. Toward effective utilization of methane: machine learning prediction of adsorption energies on metal alloys. J Phys Chem C, 2018, 122, 8315–8326.

[145] Ulissi ZW, Medford WJ, Bligaard T, Nørskov JK. To address surface reaction network complexity using scaling relations machine learning and DFT calculations. Nature Commun, 2017, 8, 14621.

[146] Wang S, Pillai HS, Xin H. Bayesian learning of chemisorption for bridging the complexity of electronic descriptors. Nature Commun, 2020, 11, 6132.

[147] Hansen HA, Shi C, Lausche AC, Peterson AA, Nørskov JK. Bifunctional alloys for the electroreduction of CO2 and CO. Phys Chem Chem Phys, 2016, 18, 9194–9201.

[148] Karamad M, Tripkovic V, Rossmeisl J. Intermetallic alloys as CO electroreduction catalysts – role of isolated active sites. ACS Catal, 2014, 4, 2268–2273.

[149] Greeley J, Stephens IEL, Bondarenko AS, Johansson TP, Hansen HA, Jaramillo TF, Rossmeisl J, Chorkendorff I, Nørskov JK. Alloys of platinum and early transition metals as oxygen reduction electrocatalysts. Nature Chem, 2009, 1, 552–556.

[150] Greeley J, Jaramillo TF, Bonde J, Chorkendorff I, Nørskov JK. Computational high-throughput screening of electrocatalytic materials for hydrogen evolution. Nature Mater, 2006, 5, 909–913.

[151] Tran K, Ulissi ZW. Active learning across intermetallics to guide discovery of electrocatalysts for CO2 reduction and H2 evolution. Nature Catal, 2018, 1, 696–703.

[152] Zhong M, Tran K, Min Y, Wang C, Wang Z, Dinh C-T, Luna PD, Yu Z, Rasouli AS, Brodersen P, Sun S, Voznyy O, Tan C-S, Askerka M, Che E, Liu M, Seifitokaldani A, Pang Y, Lo S-C, Ip A, Ulissi Z, Sargent EH. Accelerated discovery of CO2 electrocatalysts using active machine learning. Nature, 2020, 581, 178–183.

[153] Li Z, Wang S, Xin H. Toward artificial intelligence in catalysis. Nature Catal, 2018, 1, 641–642.

[154] Kitchin JR. Machine learning in catalysis. Nature Catal, 2018, 1, 230–232.

[155] Greeley J. Theoretical heterogeneous catalysis: scaling relationships and computational catalyst design. Ann Rev Chem Biomol Eng, 2016, 7, 605–635.

[156] Khorshidi A, Peterson AA. Amp: A modular approach to machine learning in atomistic simulations. Comput Phys Commun, 2016, 207, 310.

[157] Torelli DA, Francis SA, Crompton JC, Javier A, Thompson JR, Brunschwig BS, Soriaga MP, Lewis NS. Nickel–gallium-catalyzed electrochemical reduction of CO2 to highly reduced products at low overpotentials. ACS Catal, 2016, 6, 2100–2104.

[158] Lym J, Gu GH, Jung Y, Vlachos DG. Lattice convolutional neural network modeling of adsorbate coverage effects. J Phys Chem C, 2019, 123, 18951–18959.

[159] Park SW, Wolverton C. Developing an improved crystal graph convolutional neural network framework for accelerated materials discovery. Phys Rev Mater, 2020, 4, 063801.

[160] Gusarov S, Stoyanov SR, Siahrostami S. Development of fukui function based descriptors for a machine learning study of CO2 reduction. J Phys Chem C, 2020, 124, 10079–10084.

[161] Liu X, Xiao J, Peng H, Hong X, Chan K, Nørskov JK. Understanding trends in electrochemical carbon dioxide reduction Rates. Nat Commun, 2017, 8, 15438.

[162] Yang W, Morder WJ. The use of Global and local molecular parameters for the analysis of the gas-phase basicity of amines. J Am Chem Soc, 1986, 108, 5708–5711.

[163] Falivene L, Cao Z, Petta A, Serra L, Poater A, Oliva R, Scarano V, Cavallo L. Towards the online computer-aided design of catalytic pockets. Nat Chem, 2019, 11, 872–879.

[164] Liu S, Rong C, Lu T. Electronic forces as descriptors of nucleophilic and electrophilic regioselectivity and stereoselectivity. Phys Chem Chem Phys, 2017, 19, 1496–1503.

[165] Gurkan YY, Turkten N, Hatipoglu A, Cinar Z. Photocatalytic degradation of Cefazolin over N-doped TiO2 under UV and sunlight Irradiation: prediction of the reaction paths via conceptual DFT. Chem Eng J, 2012, 184, 113–124.

[166] Damoun S, Van De Woude G, Mendez F, Geerlings P. Local softness as a regioselectivity indicator in [4+2] cycloaddition reactions. J Phys Chem A, 1997, 101, 886–893.
[167] Surampudi S, Narayanan SR, Vamos E, Frank H, Halpert G, LaConti A, Kosek J, Prakash GKS, Olah GA. Advances in direct oxidation methanol fuel cells, ScienceDirect, accessed 2017-2-26.
[168] Li Z, Wang S, Chin WS, Achenie LE, Xin H. High-throughput screening of bimetallic catalysts enabled by machine learning. J Mater Chem A, 2017, 5, 24131–24138.

Steven B. Torrisi, John M. Gregoire, Junko Yano,
Matthew R. Carbone, Carla P. Gomes, Linda Hung,
Santosh K. Suram

3 Artificial intelligence for materials spectroscopy

3.1 Introduction

The materials discovery process naturally presents a slew of questions about the character of a material. These range from simply trying to learn basic facts about the atomic structure of the compound [1–3] all the way to producing a time-resolved profile of a functional process *in operando* from start to finish [4]. Spectroscopy is the process of measuring a materials' response to external electromagnetic stimulus to deduce the properties of interest. Spectroscopy can help to solve problems in experimental design and decision-making (probing response of certain energy domains, measuring particular properties), inference (moving from raw data to the property of interest), and analysis (rationalizing and interpreting the data). These are all classes of problems which artificial intelligence (AI) is well-posed to address. As such, the interface between practitioners in both spectroscopy and AI has sparked a great deal of excitement and research activity. Because characterization is a crucial process of the materials discovery pipeline, insightful pairings of algorithms and experimental protocols can lead to gains in both experimental efficiency and accuracy. These gains may significantly compress the timescales of experimental procedures and help experimentalists learn more in less time or answer questions which were previously inaccessible with a given experimental apparatus.

In this chapter, we will provide a brief survey of various spectroscopic procedures and explain how they may be understood from both a physical and algorithmic perspective. We will in turn present case studies from recent works in the field, focusing less on the scientific conclusions (which the curious reader may find in the original sources) and more so on the thought processes informing the interface between AI and spectroscopy which lead to a given approach. We hope that for new students

Steven B. Torrisi, Department of Physics, Harvard University, USA
John M. Gregoire, California Institute of Technology, USA
Junko Yano, Molecular Biophysics and Integrated Bioimaging Division, Lawrence Berkeley National Laboratory, USA
Matthew R. Carbone, Department of Chemistry, Columbia University, USA
Carla P. Gomes, Department of Computer Science, Cornell University, USA
Linda Hung, Toyota Research Institute, USA
Santosh K. Suram, Toyota Research Institute, USA

https://doi.org/10.1515/9783110738087-003

and seasoned practitioners alike, this chapter will help to equip the reader with the principles and framing that may help them solve their own research problems and conceive of new directions.

3.1.1 An algorithmic overview

The spectrum from some measurement task – whether X-ray absorption, X-ray diffraction, Raman spectroscopy, etc. – can typically be represented as a pair of length-n vectors representing n measured values at n bins (the energy of incident photons), $(E, f(E)) \in \mathbb{R}^n \times \mathbb{R}^n$ (where higher dimensional input or output can be treated as needed). There are multiple ways of representing the spectrum as a function of an incident energy f: Sometimes, the absorption is represented as a function of the photon wave vector k. These spectral datasets are prepared, featurized, and then input into an appropriate form of analysis or machine learning model. We discuss some of the details in each of these steps below.

3.1.1.1 Data preparation

Experimental datasets typically have low signal to background ratio and are also scarce, presenting a challenge for application of machine learning. In the situation where a set of spectra is collected across a smoothly varying attribute such as composition, temperature, etc., principled background subtraction can be achieved using ML methods that collectively learn the background from multiple spectra assuming that background varies smoothly as a function of the underlying attributes. In general, active approaches – wherein the subtracted background is optimized and determined on the fly, compared to the static approach where the background is pre-defined – have been demonstrated to be more efficient [5]. In addition, these approaches allow for the capture of multiple background sources, providing further gains in measurement accuracy.

In the case of scarce datasets, "data augmentation" methods are typically used, in which input data are perturbed in a way that preserves the underlying meaning of the labels, or certain data points are over-sampled during training. We will not cover them here in detail, but encourage interested readers to investigate them as they may be relevant to their individual projects.

3.1.1.2 Featurization

Application of machine learning directly on spectral data could be ineffective either due to a small dataset size or if transformations of spectral data are necessary to extract useful information. For example, consider fitting of the EXAFS equation to a measured spectrum (defined in Section 3.2.1 as equation (3.2)). Free parameters within the EXAFS equation each have a clear physical correspondence to the underlying material, and the inverse problem can be solved by numerical variation of the EXAFS equations' terms to minimize the error between the predicted and observed spectra. That is, the process of analyzing the data represents the mapping of the spectrum to the physical variables guided by the prediction of the EXAFS equation. Once these variables are obtained, due to their physics interpretation, they represent the sought "answer." The choice of featurization of the data can be thought of as a transformation when the intermediate feature space is chosen specifically to optimize interpretability or model performance. In Section 3.3.3, we will describe how the transformation of a spectrum into a polynomial-based representation can enable both enhanced interpretability as well as improved model performance.

In some cases, a predefined featurization method or a physics model is *a priori* unknown and the features that are appropriate for machine learning also need to be learned. A typical approach to learn the features/embeddings of an input spectrum are the use of encoding and decoding neural network architectures.[1] These approaches allow one to learn an embedding with an intermediate representational space of a different rank to then use as an input to a separate model, or this representation may itself be interpretable [6]. Spectral data typically suffer from variations in calibrations resulting in datasets that are theoretically equivalent but practically altered with respect to each other. In such situations, if a large number of spectra are available, convolutional neural networks may be used to learn translationally invariant and, therefore, calibration-insensitive descriptors for different materials, as was shown for manganese electron energy loss spectra [7].

3.1.1.3 Supervised vs. unsupervised methods

Machine learning algorithms typically fall into the categories of either unsupervised or supervised methods. Unsupervised methods are used to identify patterns within the data. In the case of spectroscopy, unsupervised learning can assist in preprocessing, cleaning, clustering, and preparation of data for later analysis. The case study presented in Section 3.3.1 is an example of unsupervised learning from a collection of spectra. Supervised learning describes a class of methods targeted around predicting a value of interest y_{target} for a target input dataset x_{target} (sometimes called "feature vectors") by learning a mapping function using a training dataset consisting of known input and output pairs (x_{train}, y_{train}). Our examples of supervised

---

1 Note that a valid framing of artificial neural networks is that each layer represents a transformation of the data, but sometimes in the machine learning community a particular latent space is considered as an object of study in its own right, such as in the field of generative models or the use of autoencoders.

learning are framed around extracting information about a material's state that is encoded within the spectrum.

3.1.1.4 Classification vs. regression

In the case of supervised methods, the prediction task can be framed either as classification (in which Y is a discrete set of possible classes c_i) or regression (in which Y is a d-dimensional vector of real numbers). Prediction tasks concerning the same underlying physical phenomenon can be framed as either kind. For example, consider the task of predicting the number of neighbors that a particular atomic species has using X-ray absorption spectroscopy. Within transition metal complexes, ligands surrounding a central atom will form an enclosing shape that can take on a variety of geometric shapes, such as a tetrahedron (four neighbors) or octahedron (six neighbors – see Fig. 1 of Ref. [8]). Work by Carbone *et al.* [8] treated coordination motifs as exclusive discrete classes (4-fold, 5-fold, and 6-fold coordinated), meaning that the model was configured only to bin input spectra into one of the three. Of course, this model is unable to generalize to atoms outside of those classes without the preparation of new training labels, adjustment to the model architecture, and re-training. Meanwhile, Zheng and Chen *et al.* [9] used a labeling scheme which described a central atom's coordinating environment as a vector with continuous values describing similarity to a wide set of coordination motifs. This method predicts the same target – coordination number– but via regression, instead of classification, and allows for the handling of absorbing atoms which present absorption patterns suggestive of multiple different motifs without classifying them in one configuration exclusively. The reader must decide which framing of the task is most appropriate for their use case. If less data are available, classification of the desired property into exclusive classes may be more useful to restrict the dimension of the target space, as if the output property space is high dimensional and not well sampled by the available training data then a model may have limited ability to generalize. If the goal is interpolation in a high-dimensional input space of expected spectra (such as recovering the relative concentrations of phases in a polycrystalline compound), then arrangements must be made for the collection of large quantities of experimental reference data that captures the expected variation in input spectra, or theoretical spectra which can stand in for experiments.

3.2 Spectroscopic methods: an overview

For a few forms of spectroscopy below, the relevant physical concepts will be reviewed to gain a physical understanding, and interesting examples of AI applied to respective problems in their field will be supplied. We will cover X-ray absorption spectroscopy in greater detail and briefly touch on XRD and Raman spectroscopy, highlighting some relevant use cases in each case.

3.2.1 X-ray absorption spectroscopy

Core electrons (those lying below the valence shell of electron orbitals) do not tend to participate in the bonding and chemical interaction of individual atoms with their environment. The bonding energy of these electrons to the nucleus tends to coincide with the X-ray range of electromagnetic energies (approx. 100 eV to 100+ keV). The amount of radiation which these electrons absorb under imaging can be measured, and patterns in the absorption spectra can give evidence of various structural and chemical properties of interest, ranging from elemental composition [10], oxidation number [11–15], coordination number [1, 8, 9, 16–20], ligand distance and identity [10, 11, 20], and bonding character [21]. These properties are deduced by experts using a language of "spectral fingerprints" which include peak width, peak height, peak ratios, and interpeak distance [7]. Algorithms which help to discover these fingerprints will be focused later on in Section 3.3.3.

As the energy sweeps across the electronic transition, the absorption rapidly rises, and then falls off with associated patterns of oscillation. The region of sharp increase in the absorption is known as the "edge," and occurs once the incident X-ray energy is sufficient to excite core electrons to unfilled states in the upper orbitals. The energy of peak absorption is referred to as the "white line energy" [10]. The regions just before and after the edge are referred to as the pre-edge and post-edge regions. For analysis and comparison purposes, rising edge energy has often been used. Its location can be defined as the energy value where the second derivative of the absorption crosses zero. A natural distinction arises from differences in absorption behavior depending on whether the energy of the associated X-rays is near the threshold or far above it (as high as 1 keV above [22]); the spectra associated with these two regions are known as the X-ray absorption near edge structure (XANES) and extended X-ray absorption fine structure (EXAFS). One way that XAS differentiates itself from X-ray diffraction is that XAS is useful for probing short-range (as opposed to long-range) properties in compounds of interest. Different kinds of spectra are associated with excitations from different core orbital levels; the $n = 1$ level is referred to as the K edge, the $n = 2$ level the L edge, and so on.

Absorption in the XANES region can be reckoned with in the simplest possible picture as arising from the dipole interactions between available states *via* application of Fermi's golden rule. Building off of equation (3.1) of Stern's 1974 work on the topic K-edge excitations over all final states f can be described by [23]:

$$W = \frac{2\pi^2 e^2}{\omega c^2 m^2} \sum_f \rho(E_f) \langle f | \mathbf{p} \cdot \mathbf{E} | s \rangle, \tag{3.1}$$

where $|s\rangle$ is the K-shell s orbital state, $\langle f|$ is an unoccupied p state, ρ is the density of states at the energy E_f of state f, ω is the incident X-ray frequency, $\mathbf{p}$ is the momentum operator, and $\mathbf{E}$ is the X-ray's electric field vector. Note that equation (3.1) is highly non-linear and encodes rich information about the electronic structure of a material, and so the particulars will be highly system dependent. This non-linearity helps to rationalize why machine learning has yielded so many early successes in the analysis of XANES spectra and the extraction of useful information: it is not easy to tell by inspection of this general form which particulars of individual systems will be germane for a given inverse problem, and what correlations may lie between the electronic structure encoded in the absorption pattern and any number of interesting material properties.

Meanwhile, the EXAFS equation, first stated by Sayers in 1971 (borrowing the notation of Rehr and Albers) is much more straightforward to interpret [24, 25]:

$$\chi(k) = \sum_i S_0^2 N_r \frac{f(k)}{kR^2} \sin(2kR + 2\delta_c + \Phi) e^{-2R/\lambda(k)} e^{-2\sigma^2 k^2}. \tag{3.2}$$

Here, S_0 is a amplitude factor, N_R is the coordination number of neighboring atoms at a distance R, k is the wave vector and $f(k)$ is a backscattering amplitude, δ_c and Φ are phase shifts, $\lambda(k)$ is a mean free path, and σ describes temperature-dependent broadening. The solution of the inverse problem for EXAFS data is much easier and has been in widespread use well before the advent of machine learning, as minimizing the disagreement between the EXAFS equation's predicted spectra and the measured spectra is as simple as tuning the free parameters in equation (3.2).

There are other forms of X-ray spectroscopy, which include X-ray natural circular dichroism, X-ray emission spectroscopy, resonant inelastic X-ray scattering, X-ray magnetic circular dichroism, X-ray photoemission spectroscopy and X-ray two-photon absorption [26]. We will not focus on those here, but merely mention them to highlight that other interesting and useful modalities exist.

More sophisticated forms of inverse problem solution are also appearing in the literature which deserve note; the field of applying machine learning techniques to XAS spectra goes as far back as 2002 [27] but has exploded in recent years. For example, instead of trying to learn a structural property which implicitly gives rise to a particular XANES spectrum, one approach is to use an artificial neural network which predicts model Hamiltonian parameters. This Hamiltonian is then used to recapitulate an experimentally measured spectra, and conclusions may then be drawn about the system from the Hamiltonian itself; for instance, probing the density of states of unoccupied d orbitals by measuring absorption patterns from p orbital transitions and using those to fit the physics-based model [28].

For further reading, we recommend the book *X-ray Absorption Spectroscopy for the Chemical and Materials Sciences* by Evans [10] and a review by Timoshenko and Frenkel [29].

3.2.2 X-ray diffraction

The underlying idea behind X-ray diffraction (XRD) is to exploit the scattering of X-rays by the physical arrangement of atoms to deduce interatomic spacing and therefore the configuration of atoms in a material. Incident X-rays scattering off of the periodic atoms within an irradiated system cause constructive and destructive interference and therefore a resolvable diffraction pattern. This can give clues as to the interatomic spacing and arrangement of atoms within a material. The position of peaks corresponds to unit cell dimensions, and the intensity reflects the number of atoms and their identity.

To briefly summarize the mathematical foundation of XRD, when planes of atoms are separated by a distance d, constructive interference occurs when the wavelength of light λ and the angle of scattering θ satisfy the following relationship for an integer n:

$$n\lambda = 2d\sin\theta. \tag{3.3}$$

When expressed in terms of the Bravais lattice vectors $\vec{a}_i$ and a direction of incidence $\vec{S}$, diffraction $\vec{S}_0$, wavelengths λ, and lattice planes (h, k, l), the Bragg condition can be written [30]:

$$\frac{\vec{S} - \vec{S}_0}{\lambda} = \vec{d}_{hkl}^* = h\vec{a}_1 + k\vec{a}_2 + l\vec{a}_3. \tag{3.4}$$

Typically, the pattern can be measured by rotating a detector around an irradiated sample and measuring the scattered X-rays at each individual angle. For crystalline materials, the expected spectra will involve sharp delta function-like peaks, and for amorphous materials, it will be a smooth curve due to the breakdown of long-range ordering. Similarly, to know how the absorption pattern shown by XAS will arise from the sum of the contributions from each absorbing atom, XRD spectra will integrate the contributions from each individual phase within a measured sample.

Various factors may impact the equilibrium lattice spacing and induce local disorder, thereby affecting the XRD spectra and introducing noise. For instance, dopants, defects, stress, and temperature variation will all serve to shift, broaden, or suppress diffraction peaks.

An idealized picture of a material typically involves a single crystal phase, but XRD finds application in polycrystalline and amorphous substances as well [30]. Polycrystalline phases have grain boundaries within the material in which perfect periodicity is broken and crystalline phases intersect at random angles with different terminations. XRD applied to materials in powdered forms take into account this random distribution of orientations and attempt to describe the coinciding underlying phases. The combination of deep learning techniques with simulated data for XRD holds great promise for multiphase mixtures. For instance, the generation of a library of reference phases and associated XRD patterns which can be combined in many linear combinations to create a synthetic "multiphase" library to train a neural network, which then can be used to identify the constituent combination of phases in a synthetic spectrum [31]. Another recent example showed how an iterative loop combining XRD with density functional theory to determine the criteria for stability of the cubic phase of a perovskite as a function of cation doping [32].

3.2.3 Raman spectroscopy

Raman spectroscopy characterizes a molecule or material by measuring the inelastic scattering of electromagnetic radiation with individual degrees of freedom. This coupling is measured by detecting a change in frequency which comes from a scattering event. These degrees of freedom can include ro-vibrational motion as well as electronic excitations and electron-plasma oscillations in a molecule or material of interest [33, 34]. The particular frequencies of resonance yield a spectral "fingerprint", as similar to infrared spectroscopy, but with a different selection rule. This fingerprint can be used in conjunction with other methods, such as first-principles calculations, to deduce what bonding configurations gave rise to an imaged spectrum, thus revealing facts about the underlying arrangement of atoms.

Some interesting applications of AI to Raman spectroscopy have been the use of support vector machines in tandem with measured spectra to accurately construct the phase diagram of a ferroelectric material [35]. Convolutional neural networks also offer advantages for Raman spectra classification due to their ability to detect local correlations between spectral features, and were shown to ameliorate the need for pre-processing in imaging biological data [36].

3.3 Case studies

Given the broad range of spectroscopy techniques and the breadth in material/chemical properties they aim to characterize, we present three case studies which span different techniques, stages of the scientific process, and materials systems.

We start with the broadly applicable topic of background modeling and demonstrate how AI can identify background signals with superhuman analytical capabilities in Raman spectroscopy data in Section 3.3.1. Background subtraction is typically considered a data-processing step, although its importance in data analysis motivates its elevation to a data-interpretation step wherein an AI algorithm makes probabilistic inferences concerning the existence and extent of background signal. Phase mapping is an even more challenging data interpretation task where some datasets confound human experts to the point that no reasonable solution can be obtained without assistance from AI. In Section 3.3.2, we consider the case study of inferring the underlying phase diagram characterized by a collection of XRD patterns from different compositions in a three-component composition space. Here, different datasets pose challenges related to complexity where so many features need to be simultaneously considered that identification of phases in each measured XRD pattern can only be achieved by incorporating prior knowledge; in particular, thermodynamic rules that describe a valid phase diagram. This case study also affords a discussion of identification and characterization of datasets that can be interpreted in different ways that are equally valid and consistent with the experiments. AI exploration of *all* valid solutions is more challenging than identification of *one* valid solution, which should always be considered in development of AI for spectroscopy.

Afterward, we will discuss the solution of inverse problems in X-ray absorption spectroscopy. We will explicate in Section 3.3.3 the question of drawing correspondences between the structural or electronic properties which are encoded within an XAS spectrum via random forests and the appropriate featurization of the input spectrum. This approach demonstrates how featurization combined with an interpretable model allows for insights to be extracted from a spectrum where the correspondence between spectra features and a property of interest is not *a priori* known. The discovery of these fingerprints helps expand the frontier of expert knowledge, and can also help build the trustworthiness of models which will see integration into experimental workflows. This trust can come from not only verifying if the model is fitting to meaningful parts of the input signal, but also in predicting what kinds of input data will be outside of the training dataset and are therefore at higher risk of returning inaccurate results.

3.3.1 Modeling background signals in spectroscopic data

As our introductory case study, we consider background removal as an exemplar of spectroscopic data analysis enhanced by AI. In particular, we will highlight the multicomponent background learning (MCBL) algorithm reported by Ament et al. [37].

Information extraction from noisy and background-containing signals is pervasive in physical sciences research [38–42]. Distinguishing the signal of interest from background signals comprises a major hurdle, and any errors in making these distinctions can alter data interpretation. A commonality of all spectroscopy techniques is that there is a sample of interest whose spectral signal is intended to be measured, yet the measured spectrum contains signal that is not related to the sample of interest, or at least the relationship to the sample is not the intended one.

The signal that is not related to the sample can be considered as "background signal," although the most challenging cases of background removal are when a signal is present that is related to the sample of interest but the relation is due to physics that is beyond the intended spectroscopy. Examples of such signals in X-ray diffraction include fluorescence of the sample or attenuation of a background X-ray signal by the sample. In Raman spectroscopy, one such example is a Raman signal from the substrate that is attenuated by the sample, where the level of attenuation is determined by the absorption coefficient of the sample of interest at the photon energy of the excitation laser (a physical property that is markedly different from the Raman physics targeted by the spectroscopy experiment). When the underlying physics of the background signal are known, a parameterized model for its application toward a given measured spectrum may be derived, and determination of the parameters via regression provides a principled model for the removal of that signal from the spectrum. Utilization of this approach has to be considered in the context of the specific spectroscopy experiment at hand, and for the present case study, we consider a generalized version of this approach called multicomponent background learning wherein we can impose some knowledge about the background signals without a closed-form expression of the background signals.

To conceptualize an algorithm for automated background modeling we can first consider how humans perform this task, where the specific method varies among spectroscopic techniques although one commonality is that a human uses their experience from having seen many measured spectra from other samples. Combined with an expectation of what the signal from their sample should look like, they can adjust a background model to remove component signals of the spectrum that they recognize as occurring in many or all measurements of other samples, producing a background-removed spectrum that matches their expectations. The reasons to automate this process include the practical desire to mitigate the expense of human involvement in data analysis, but perhaps more critically, an automated framework for assessing background can remove human bias and variability in background modeling. In some sense, machine learning for background modeling seeks to elevate background modeling from a manual art to an automated science.

As noted above, if there is a known functional form for the background signal and there is only one source of background, traditional regression approaches may be sufficient. When there are multiple and/or an unknown number of background sources that can each vary as a function of time and/or depend on the properties of the material being measured, we require the background modeling to learn the shape of the background signal(s). Returning to the above example of a sample-specific attenuation of a background signal in Raman spectroscopy, an additional concurrent background signal may depend on the ambient temperature of the measurement that is variable but not measured. These multiple background sources have signals that can vary independently of each other, motivating explicit modeling of multiple background signals in each spectrum.

For the present use case, we consider the situation in which there are multiple sources of background signal that have a characteristic spectral shape but whose intensities can independently vary over the set of measured spectra. Learning the shape and relative intensities of an arbitrary number of background signals is a challenging machine learning task that MCBL tackles by making an additional assumption that as many spectra are acquired, the background signals will appear many times while any single signal of interest will appear in a much smaller fraction of the total number of measured spectra. This assumption wouldn't apply to a situation in which the same sample is being measured repeatedly, as that series of spectra does not allow identification of which signals arise from the sample of interest as opposed to from background sources. Typically, a collection of samples are being measured whose spectral properties are distinct in some way. In this case, the spectral signal from one sample will not have the same shape as the spectral signal from other samples, while background sources maintain their shape.

As an example dataset, consider Raman spectroscopy acquired on 2121 metal oxide compositions that include all possible mixtures of the cation elements Mn, Fe, Ni, Cu, Co, and Zn with 10 atom% composition intervals and 1 to 4 of these cation elements appearing in each sample. The set of 2121 Raman spectra will include many examples of absence and presence of each element, including many combinations with different cation elements. As a result, a Raman mode related to a given cation element and its incorporation in a crystal structure will only be present in a small fraction of the measured spectra. The strategy is thus to model background signals as the signals that appear across many measured spectra, although this task is challenging when the background signals have highly variable intensities or have special features that overlap those of some or all of the samples of interest.

Returning to the algorithmic thinking of a human analyst, the assessed probability that a given portion of a measured spectrum is due to a background source would be related to how often a similar signal is observed in measured spectra from other samples. MCBL formalizes this concept in a probabilistic framework in which each measured spectrum is decomposed as a sum of component signals, one from the sample of interest and any number of signals from background sources. To establish an appropriate framework, consider that a successful background model would produce a background signal for each spectrum where after background removal, the net spectrum contains a combination of the signal of interest and experimental noise that is not systematically related to any background source. The goal is thus to identify a background model that meets this goal, which is achieved by maximizing the likelihood that the background model leads to a background-subtracted spectrum exhibiting these characteristics. The experimental noise that is not related to a specific background source can generally be considered to produce intensities from a Gaussian (normal) distribution, $\mathcal{N}_{\mu,\sigma}$. For Raman spectroscopy, the signal of interest is derived from Raman modes within the material that gives rise to peaks in the measured pattern. As a result, after background removal, when signal of interest (a peak) is present, the distribution of measured intensities will not follow a Gaussian distribution. A more accurate probability distribution for the measured intensities of peaks is the exponentially modified Gaussian (EMG) [43, 44]. To model multiple background signals with consistent shapes but variable intensities, matrix factorization can be deployed where each basis component corresponds to the spectral shape of a background signal whose activation (relative intensity) in a given measured spectrum varies among the set of spectra under consideration. The basis patterns and their activation matrix comprise a host of parameters that can be determined alongside the parameters of the N and the EMG distributions of equation (3.5) to maximize the likelihood that the non-background intensities follow these distributions when the signal of interest is absent and present, respectively. Defining R_{ij} to be the intensity that remains in data point j from measured spectrum i after background removal, the general expression for the distribution of R_{ij} is

$$\mathrm{EMG}_{\mu, \sigma, \lambda, Z_{ij}}\left(R_{ij}\right) = \left(1 - Z_{ij}\right)\mathcal{N}_{\mu, \sigma}\left(R_{ij}\right) + Z_{ij}\mathrm{EMG}_{\mu, \sigma, \lambda}\left(R_{ij}\right), \tag{3.5}$$

where Z_{ij} indicates whether signal of interest is absent ($Z_{ij} = 0$) or present ($Z_{ij} = 1$) in R_{ij}. Maximizing the joint likelihood over all data points in all spectra given this expression for the probability density provides a measure of the probability that the signal of interest is present in each data point (Z_{ij}) as well as the shape of each background signal, the overall background signal in each measured spectrum, and thus the net spectrum of interest after background removal. A full implementation of this algorithm was previously described elsewhere [37].

Illustrative results from MCBL background modeling on the set of 2121 Raman spectra are shown in Fig. 3.1. Figure 3.1a contains a particularly challenging background signal in which a series of Raman peaks from the substrate overlap with much weaker Raman peaks from the sample of interest. The results from traditional background-removal algorithms, polynomial baseline models, and the Sonneveld–Visser algorithm of Ref. [38] are shown in Fig. 3.1b. Since these algorithms model the background of each spectrum independently, identification of the background signal in Fig. 3.1a is untenable. Figure 3.1a also shows the results of the background modeling where only one background source is considered, which still outperforms traditional methods but does not sufficiently remove background signal. Regularization of the background intensities in MCBL enables modeling of a large number of background sources without overfitting the background. By choosing an upper limit for the number of background sources, in this case 16, the background model can be learned with the only human decisions being the choice of the probability distributions N and EMG. Using this 16-source model, select Raman signals from the collection from 2121 are shown in Fig. 3.1c, and the Z_{ij} is used to model the probability that a signal of interest is present in each individual data point, as shown in Fig. 3.1d. The resulting background-removed signals are shown in Fig. 3.1d.

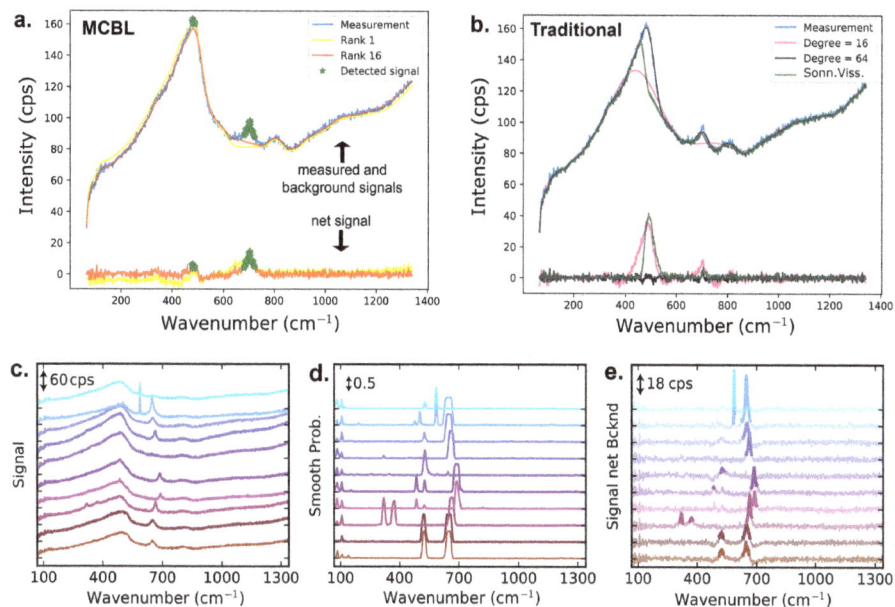

Fig. 3.1: a. A single Raman spectrum is shown along with its background model learned by the MCBL algorithm from a collection of 2121 Raman spectra. The MCBL model when considering a single (Rank 1) and 16 (Rank 16) background source(s) is shown, along with the background-subtracted signals where individual data points determined to have a greater than 50% probability of containing intensity from the sample of interested are denoted in green. b. Background modeling of the same spectrum under-estimates the background signal when using low-order polynomials, and increasing the polynomial order to model the background results in removal of the signal of interest. c. A series of 10 representative Raman spectra from the collection of 2121. d. The probability that each measured data point contains signal from the sample of interest, as determined by the MCBL model. e. The spectra after background removal, with partial transparency for data points that have less than 50% probability of containing signal from the sample. This figure was adapted from Figs. 2 and 3 from Ament et al. [37].

3.3.2 Phase mapping from XRD spectra

For our next case study, we consider the problem of phase mapping wherein the ultimate goal is to construct a phase diagram from a collection of measured XRD patterns from a given composition space [45]. The key analytical step toward this goal is to identify what phase(s) are present in each measured XRD pattern. Considering each phase to be a source of XRD signal, materials that are not phase pure will have an XRD pattern that contains a mixture of these signals. Identifying the source signals and their relative intensities constitutes a demixing or source-separation problem, which motivates use of source-separation techniques to perform phase mapping [46–50]. Clustering techniques have also been developed [51–53] although they are not the focus of the present case study. Phase mapping is a long-standing challenge in exploration of materials containing more than two components, prompting our focus on three-component systems in the present case study. The results of phase mapping provide a composition map of the molar fraction of each phase and, ideally, the identity of each phase, which can be used to construct a phase diagram and/or establish composition–structure–property relationships, as was recently demonstrated with the use of a phase mapping result as a template for active learning-based property optimization [54]. In this case study, we focus on the recently reported algorithms agile factor decomposition (AgileFD) [48, 49] and CRYSTAL [50]. As described further below, these algorithms employ "constraint reasoning," a family of AI methods that enforce rules about a meaningful solution for a given task, where the constraints are implemented within the solver and a search method is used as the inference procedure. Rules concerning valid solutions are ubiquitous in analysis of scientific data, making development of solvers for constraint-reasoning problems a historic and active area of research [55–57].

Among the algorithms for tackling source-separations problems, matrix factorization is particularly intuitive, where one matrix contains the individual XRD patterns for each phase ("basis patterns"), and another matrix contains the weights that govern the mixing of these phase-pure patterns in the measured patterns that typically contain more than one phase (see Fig. 3.2). Since both the XRD intensity and the amount of each phase in a sample are non-negative numbers, this problem corresponds to a specific type of matrix factorization called non-negative matrix factorization (NMF). Indeed, some phase-mapping datasets can be readily solved by NMF [46, 47], although importantly such source-separation algorithms can be considered to be data-compression techniques wherein a large matrix of measured XRD patterns is factored into two much smaller matrices.

One source of incompatibility of NMF with phase mapping is underscored by the fact that we do not seek the most compressed representation of the data but rather a compressed version that meets specific requirements. Two such requirements are that (i) the component signals correspond to phase-pure XRD patterns and (ii) the activation of the phases is physically meaningful. The former requirement is challenging to meet when the dataset contains multiple phases that appear together more than they appear separately, which drives the matrix factorization to use a basis pattern that is some combination of the multiple phases, violating requirement (i). The latter requirement is more subtle and can be challenged by a number of experimental factors. Considering the high-level goal of constructing a phase diagram, the activation of each phase should vary systematically in the composition graph of the composition space under consideration, which is an attribute of the solution for which standard NMF is agnostic. Both of these requirements are challenged

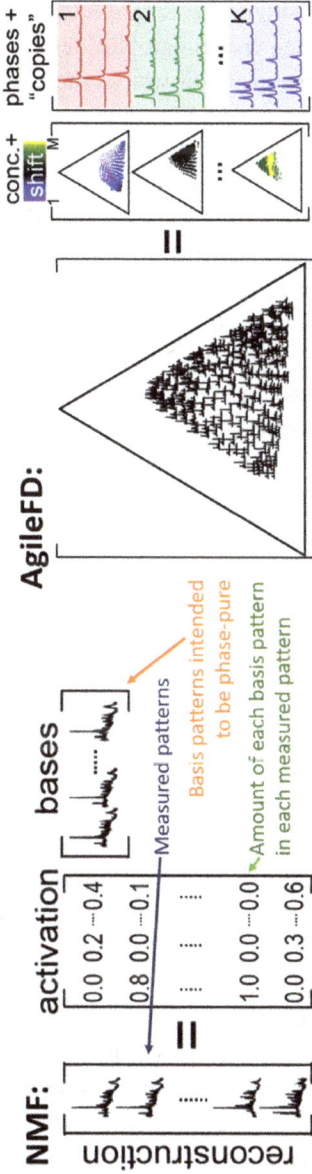

Fig. 3.2: Illustration of phase mapping as matrix factorization where a matrix of measured XRD patterns is factored into two matrices, one containing the basis patterns that are intended to be phase-pure and another containing the amount of each basis pattern in each measure pattern. AgileFD extends this concept by including multiple copies of each basis pattern to model alloying-based multiplicative peak shifting. While NMF has no inherent representation for the composition corresponding to each XRD pattern, the AgileFD illustration shows the XRD patterns at their respective location within the ternary composition graph, where the composition map of phase activation is used to apply thermodynamic constraints. Right-hand side is adapted from Fig. 3 of Bai et al. [58].

by additional experimental realities, primarily those arising from alloying. Alloying involves composition-dependent speciation of at least one crystallographic site in a phase, which can alter the relative intensities of diffraction peaks as well as the peak positions due to modulation of the lattice constants. As a result, XRD patterns from different alloy compositions of the same phase will have different patterns but constitute the same phase in the phase diagram, which runs contrary to the underlying assumption of NMF that each source has the same signal in each measurement. Other alterations to source signals include changes in grain size that influence peak width or changes in grain orientation that results in alterations to relative peak intensities. Collectively, these challenges can be considered as an additional requirement for the phase-mapping solution: (iii) versions of the source signals that are modified according to alloying, changes in crystallite size, etc. should be considered to be the same phase. In total, NMF is not guaranteed to meet requirements (i) or (ii) and is largely incompatible with requirement (iii). While standard NMF is thus ill-suited to the problem, we can use the NMF construction to consider algorithmic advancements that facilitate compliance with the three primary requirements.

A modification to NMF that would promote satisfaction of requirement (i) would be to compare each de-mixed pattern to patterns of known pure-phase materials and derive a loss based on the difference between each de-mixed basis pattern and its most closely matched phase-pure pattern. While conceptually straightforward, challenges with implementing this strategy include the computational expense of comparing each basis pattern to potentially hundreds of known phase-pure materials in the composition system. AgileFD and CRYSTAL use this concept to evaluate, but not derive solutions. The strategy is to identify a valid solution with respect to the other requirements and afterward confirm that requirement (i) is met as well.

Requirement (ii) concerns the activations of basis patterns. Encouraging smoothness in the compositional variation in activations could be achieved by deriving a loss function that considers the compositions of each row of the activation matrix. Importantly, traditional NMF has no capacity to consider the compositions but a custom loss function could be used to look up the composition based on matrix index and calculate properties of the compositional variation in activation of each phase. It is not immediately clear what loss function would be optimal for enforcing the requirement because the known rules of phase diagrams cannot be directly expressed as a differentiable function of activation vs. composition. The rules are based on thermodynamics, and we will describe them assuming thermodynamic equilibrium, although parameterized versions of these rules can allow deviations from thermodynamic equilibrium in solutions where the extent of deviation is chosen by the user.

For example, the most straightforward thermodynamic rule is Gibbs' phase rule that limits the number of coexisting phases to be no more than the number of variable components in the composition space, for example a maximum of three phases for a three-component composition space. A user may want to extend to this rule to allow for four co-existing phases when they expect that their materials are not in thermodynamic equilibrium, making the desired algorithmic extension of NMF to enforce that each XRD pattern contain no more than G phases, or source patterns, where G is selected by the user. Conceptually, this thermodynamic rule could be promoted through regularization on the activation matrix, although such an approach does not enable enforcement of other thermodynamic rules of interest. For example, alloying in the local composition region comprises a thermodynamic degree of freedom and lowers the Gibbs' phase rule maximum for the number of co-existing phases. Since this rule inherently involves detection of alloying, we will return to it when considering requirement (iii). Also, and perhaps most importantly, for being able to construct a meaningful phase diagram from the NMF solution, the activation of each phase must form a single connected region of the composition graph. Similarly, each combination of phases that exists in the data should form a connected region of the composition graph. Collectively, these thermodynamic requirements correspond to combinatorial constraints in which the validity of the solution for one XRD pattern depends on the solution of many other XRD patterns. Explicitly enforcing such constraints cannot practically be achieved through modification of the NMF loss function, prompting a strategy in which the matrix factorization is interleaved with a constraint-reasoning algorithm that corrects the solution to obey the thermodynamic rules. After imposing the rules, the corrected solution will likely no longer be optimal with respect to the NMF reconstruction loss, prompting further NMF optimization. In practice, solution convergence is optimized by interleaving optimization of the source separation with application of constraints, as demonstrated by interleaved AgileFD (IAFD), the source-separation engine of CRYSTAL [50] shown in Fig. 3.3. Implementing this approach can be challenging as a competition between NMF optimization and constraint satisfaction can cause the solver to become stuck in a local minimum, which CRYSTAL addresses with a combined strategy of gradual constraint enforcement and random restarts.

The extension of NMF to incorporate constraint reasoning enables enforcement of the thermodynamic rules in the solutions, although the Gibbs' rule cannot be implemented until alloying is explicitly modeled, which brings us to requirement (iii). The variations in peak position and peak width pose the greatest challenges to matrix factorization because there can be measured intensity for a phase where the source model has none or vice versa. Variations in peak intensity can be explicitly modeled or be implicitly allowed by choice of reconstruction loss function. For example, the KL divergence loss function strongly penalizes erroneous modeling of the absence or presence of intensity compared to reconstruction error wherein the modeled and measured intensity are both large but not equal. By using this loss function for NMF, the extension of NMF in AgileFD focuses on modeling variations in peak width and peak positions. The alloying-based modification in peak positions corresponds to a single phase appearing as multiple modified versions in the set of measured XRD patterns. From known phase diagrams, we know that lattice

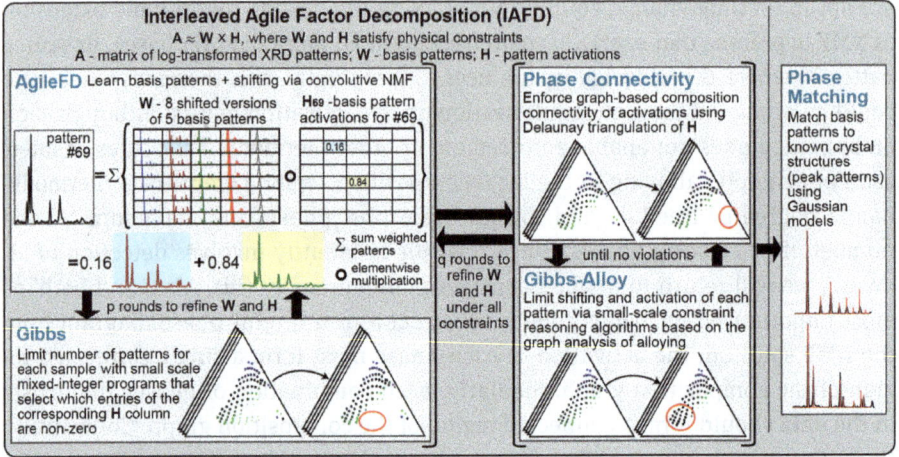

Fig. 3.3: The IAFD bot network solves phase mapping as a constrained matrix factorization problem in which the input XRD pattern matrix (A) is decomposed into factors W and H such that $W \times H$ approximates A while satisfying physical constraints. W encodes the characteristic patterns of pure crystal phases (including shifted versions) and H their activations, which dictate both the amount and the pattern shifting extent of each pure phase in each XRD measurement. IAFD starts with p (typically 3) rounds of interactions between the AgileFD and Gibbs bots followed by rounds of iterations between the Gibbs alloy and phase connectivity bots, until all the constraints are satisfied. AgileFD performs matrix factorization using light-weight multiplicative updating rules, without enforcing the combinatorial physical constraints. The AgileFD solution's violations of the connectivity constraint and the constraints based on Gibbs' phase rule are repaired by the corresponding bots in an interleaved manner using efficient algorithms (red circles highlight repaired activations of H). The entire procedure is repeated for solution refinement (typically q = 2), and the resulting generated basis patterns are passed to the phase-matching bot to identify the crystal structures by comparison with known phases. The figure illustrates a representative XRD pattern that is decomposed into shifted versions of two different basis patterns (0.16 and 0.84 of each respectively). This figure is adopted from Fig. 3 of Gomes et al. [50].

constants do not typically change by more than 10% due to alloying and that alloying-based changes to lattice constant can typically be approximated as an isotropic lattice expansion or contraction. In such an approximation, the scattering vector magnitude of each peak shifts by a multiplicative factor, and under the isotropic approximation, all peaks from the phase shift by the same multiplicative factor for an alloy of a given composition. This is a very specific type of modification to a phase-pure pattern that can be modeled in NMF by considering multiple shifted versions of each basis pattern. The typically observed peak width can be used to identify the number of shifted versions of a phase that need to be considered.

For example, if the minimum peak width is typically 2% of the scattering vector magnitude, then consider the alloy version of a phase that has the highest average lattice constant, which corresponds to the lowest scattering vector magnitudes of the peaks in the phase. Given the 10% limit on multiplicative peak shifting, modifications of this phase pattern with peaks shifted by multiplicative factors of 2%, 4%, 6%, 8%, and 10% means that any alloyed variant of that phase will have an XRD pattern that strongly overlaps with one or up to a few of these shifted variants, depending on the peak width in the measured XRD pattern of the alloy. In this example, there are six versions of each phase's pattern, which in the NMF construction can be modeled as $6 \times K$ rows in the basis pattern matrix to model K phases. In this case, each set of six rows is not treated as independent patterns (as would be the case in standard NMF), but mutually constrained to be shifted versions of each other. Allowing multiple versions to be activated for a given measured pattern simultaneously addresses variability in peak width as wider peaks can be modeled by partial activation of different versions of the pattern with shifted peaks. In practice, this motivates using a larger number of shifted versions of each phase.

Concerning the detection of alloying, if alloying causes a pattern shift for the instantiation of a given phase in two neighboring compositions, the corresponding NMF solution will contain a change in the distribution of shifted variants of the source patterns that are activated. Detection of this composition-dependent lattice constant can be added to the constraint-reasoning module that enforces thermodynamic rules. The thermodynamic rules based on activation also need to consider the total activation over all shifted variants, which is also readily done in the constraint-reasoning module since its execution is interleaved with NMF optimization.

In practice, these extensions of NMF and use of constraint reasoning to enforce thermodynamic rules have been successfully deployed for solving complex phase-mapping problems such as the V–Mn–Nb oxide system wherein eight phases and their alloy variants were automatically identified from a collection of XRD patterns (Fig. 3.4). In that case, the compositionally tuned alloy of one of the phases gave rise to a systematic change in band gap energy (Fig. 3.4c), demonstrating that not only is explicit modeling of alloying required to solve complex phase-mapping datasets but also that the resulting characterization of alloying in new composition spaces provides opportunities for discovery of materials with tunable properties.

While IAFD by definition provides solutions that obey thermodynamic rules, its limitations have been observed in two ways, both related to the analysis step noted above wherein the identification of the phase corresponding to each demixed pattern is performed as a post-processing step. When a phase-mapping dataset violates the assumption that each phase appears as different mixtures with other phases, the IAFD solution may not have sufficient data to demix phase-pure basis patterns. In this case, at least one basis pattern may have no readily identifiable phase. The measured patterns that employ that non-phase-pure basis pattern will also be incorrectly modeled, which can compromise the fidelity of other basis patterns. In practice, identification of the violated assumption can be corrected by providing manual guidance or constraints to the IAFD solver, although a model that circumvents such limitations is being actively developed.

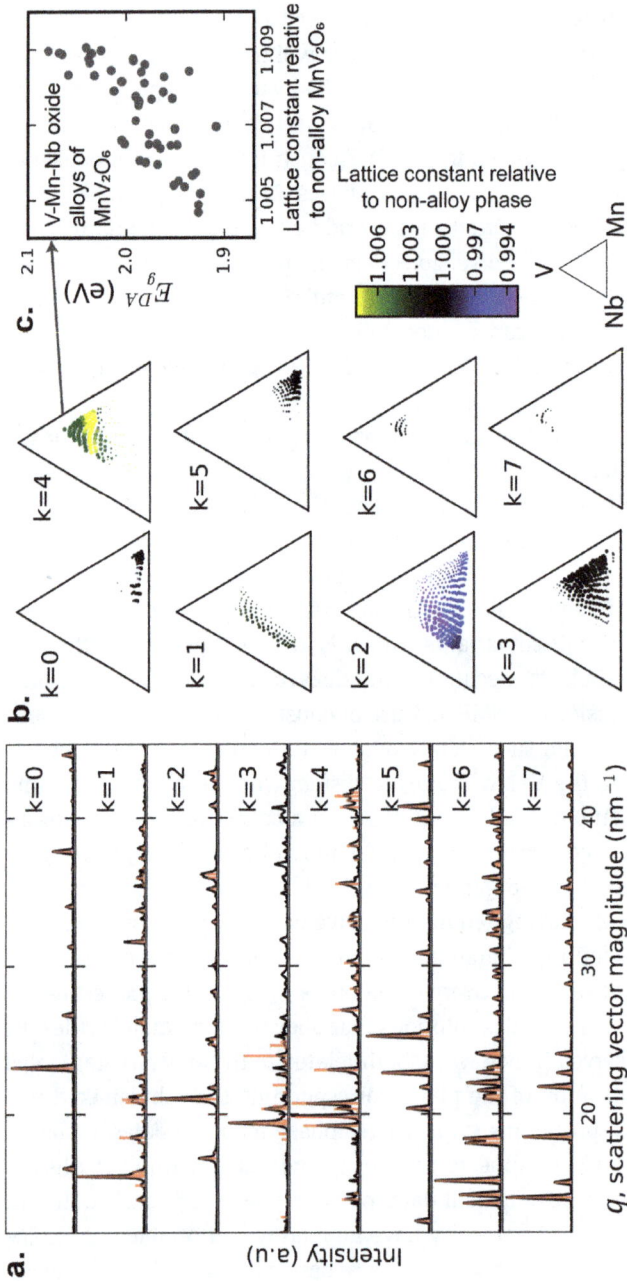

Fig. 3.4: Illustration of phase mapping results from the AgileFD algorithm wherein XRD patterns from 317 unique V–Mn–Nb oxide samples were collectively analyzed by factoring each measured pattern into no more than 3 of the $K = 8$ basis patterns (or their alloy-shifted variants). a. The eight basis patterns and their match to known phases (red). b. The composition map of activation (point size) and lattice constant (color scale at bottom right) relative to the matched phase. c. The $k = 4$ phase is the MnV_2O_6 phase that appears in high phase purity with a range of different compositions and lattice constants. The band gap was separately measured and compared to the lattice constant, showing that alloy composition alters the lattice constant and enables band gap tuning over a range critical to solar energy technologies. This figure is adopted from Suram et al. [49].

The second type of issue that can arise when matching phases to demixed patterns is that multiple different phases may match approximately equally well, meaning that the demixed pattern does not uniquely identify the phase that is present. Furthermore, repeating the analysis with different random initialization can lead to basis patterns that are best matched to different known phases than the first solution. In practice, this limitation has been determined to be primarily due to the limitations of the dataset, as opposed to the algorithm, although that finding was only enabled through our development of an AI system to explore the solution space.

A primary benefit of IAFD over prior phase-mapping algorithms is its orders-of-magnitude reduction in computational cost, which enables liberal exploration of the variety of solutions that can be obtained with different random initializations and parameters. Automating this process involved integration of the NMF and constraint-reasoning components of IAFD with the phase matching and other ancillary processes, collectively creating the multi-agent AI system CRYSTAL [50]. For a single phase-mapping dataset from the Pd–Rh–Ta composition system, 2,500 candidate phase diagrams were automatically generated and compared to identify several unique solutions involving different phases and different physical interpretation of the dataset while obeying thermodynamic rules and providing a nearly equivalent reconstruction of the measured data. The explanation for this occurrence is fairly straightforward: the nanocrystalline grains resulted in broad XRD peaks, and combined with alloying and known phases having strongly overlapping signals, the source data was insufficient to uniquely identify all containing present phases. While CRYSTAL's exploration of the solution space identified that additional data would be required to fully solve the phase diagram, more critically for the broader picture of AI for spectroscopy is the demonstration that some datasets have multiple valid interpretations. In this case study, and more generally, an AI method seeks to automate analysis of the data, and if the result seems reasonable and obeys all expectations (for example, thermodynamic rules in phase mapping), the user will naturally accept that answer as the conclusion and proceed to employ it in their research. Exploring the stability of solutions and the implications of alternative explanations of measured data is an often overlooked but critical aspect of spectral data analysis, for which CRYSTAL provides one approach.

The CRYSTAL AI framework provides an important demonstration of a multi-agent approach to solving complex problems in AI for spectroscopy. Here, an "agent" is an algorithmic bot that can perform one component of data processing and interpretation. This modular approach to AI algorithm development enables the overall AI system, CRYSTAL in this case, to coordinate activities by summoning the expertise of individual agents as needed. This approach is analogous to a collaborative team of experts wherein a project leader summons experts as needed through the course of research. The analogy can be extended to volunteered contributions from experts in the collaborative team setting, where individual agents run in passive monitoring mode and "speak up" when they see an opportunity to contribute. With the growing purview of AI for spectroscopy and beyond, the research tasks will increasingly not fit into the traditional concept of a single machine learning model, motivating further adoption of the multi-agent approach.

3.3.3 ML for XANES case study: tuning feature space for interpretability

Interpretation of X-ray absorption spectra hinges on the availability of human-discernible "fingerprints" in the spectra which may be mapped back to a property of interest. In this section, we will explain and show step by step the sequence of reasoning which lead to the development of multiscale polynomial featurization as an approach for developing interpretable random forest based models for the analysis of XANES spectra. The problem will be worked through using toy model spectra which have a straightforward functional form to demonstrate the power and utility of this method for extracting information from spectra where the relationship of cause and effect between a desired property and the spectral curve is not *a priori* known.

3.3.3.1 Setting up the problem

Interpretability in machine learning algorithms is an active field of research with interesting philosophical underpinnings [59]. What does it mean for a model to be interpretable? The EXAFS equation may be considered "interpretable" because all of the free parameters in the equation each have easily graspable physical correspondences. In other words, the degrees of freedom in the function can be related back to a physical picture of the system in question. A non-parametric black-box model which attempts to solve an inverse characterization problem by mapping from spectra to some output value "in the blind", without any underlying physical model built-in, may be able to provide interpretability by revealing which parts of the domain are most germane to the regression task; the job of connecting those features to physical properties must then be left to the scientist.

However, even highlighting *which* parts of the spectrum are most important for the model's output prediction may be valuable. For example, if the model is observed to be working *via* indicators which have a readily understandable physical correspondence, the model may be more expected to generalize to new data. If the practitioner does not have a strong intuition for a system in question, strong trends in the feature space highlighted by the model may accelerate the process of gaining familiarity with the compound under study.

Some models like random forests [60] make it easier to interpret which parts of the input space are most germane to the regression task. For random forests in particular, the relative importance of one feature or another can be assigned a quantitative weight based on the contribution to the loss function that the feature brings (see Ref. [60] for an overview on random forests). If an entire XANES spectrum was input into a random forest model, where each individual value of absorption is a feature, then the importance of different parts of the spectral domain could be determined. However, this may make it difficult to understand the role that correlations between different points play: for example, if three separate values at adjacent energies E_0, E_1, and E_2 were all indicated as important, are each values being used in the same 'paths' down individual decision trees? Not necessarily, due to the random nature of selecting which features to use for successively added leaves. Do the successive "paths" down a tree test for the presence of a linear, or quadratic-like trend in the spectra across the three points? We can remove the guesswork if the features are constructed in such a way that they are easily interpretable, or can capture features of a spectrum beyond the importance of individual domain regions, then relevant features of the spectrum can be teased out. If these features have a clear mathematical or physical relevance, then it makes the task of verifying if the model agrees with the physical picture – or if it points the way toward new insights – easier.

One such technique by Torrisi *et al.* [20] allows a researcher to featurize the spectra in a way as to reveal the relevant domains which are used by a random forest model in solving an inverse problem. This idea came from asking a motivating question: different regions of energy space correspond to different underlying physical causes in XAS spectra. For instance, the pre-edge region may contain weak transitions due to mixing of dipole transition allowed and dipole-forbidden orbitals [17], or interactions with the unfilled orbitals in neighboring ligands. The shift of the edge itself corresponds to a change in binding energy in the core electrons which are being excited. Because we know that certain regions of the domain have this physical correspondence, it stands to reason that for a particular prediction task, we expect certain features to be more relevant.

One prominent example is that in titanium complexes, the pre-edge region has features which are suppressed by increasing coordination number. Therefore, in order to discern what is present, an expert may only need to look at the pre-edge region. However, the question remains as to how *new* spectrum-property correspondences may be discovered. Can they be determined for various prediction tasks of interest without prior knowledge?

3.3.3.2 Multiscale featurization

We can carry out the task of featurizing the spectrum in a few ways. First, one can use a package like Numpy [61] to ensure that all spectra are formatted on the same grid (using an interpolating function). Next, ranges of interest can be picked out and fit using a polynomial function of varying order (to avoid overfitting, we suggest using low-order polynomials). The coefficients of these polynomials must then be fed into the random forest model as a concatenation of multiple vectors corresponding to different polynomials i and their respective coefficients a of order n as $\{a_{0,i}, a_{1,i}, a_{2,i}, a_{3,i}\}$.

This will then allow the user to make predictions and later to rank the importance of individual vectors. In this example, we will reproduce the work from Torrisi *et al.* [20] in which multiscale polynomial featurization was demonstrated to provide insights into what features of interest are used in an algorithm's prediction.

We can demonstrate this workflow by example for a set of "fake" spectra. These are meant to superficially resemble XANES spectra, originating from a functional form which makes it easy to see the "hidden variables" that give rise to the resultant shape of the curve. We'll henceforth refer to these curves that we generate as "spectra" by analogy, even if they do not represent any real physical simulation or process, to keep the exposition easy to follow.

Supposing that some feature a which we seek to measure has a straightforward correspondence with the sharpness of the edge. We generate the "fake" spectra using the following function as the sum of a sigmoid (to model the absence of signal in the pre-edge, and saturation into the edge and post-edge region) and two Gaussians in the post-edge; one positive, and one negative to create a peak and a trough. This curve will be parameterized by six variables a, b, c, d, e, f:

$$\mu(E) = \frac{1}{1 + \exp[-(3a)(E+b)]} + \frac{1}{d\sqrt{2\pi}} \exp\left[-\frac{(E-c)^2}{2d^2}\right] - \frac{1}{f\sqrt{2\pi}} \exp\left[-\frac{(E-e)^2}{2f^2}\right]$$

(3.6)

a corresponds to the "sharpness" of the edge, b corresponds to a shift in the edge, c and e are locations of post-edge peaks, and d and f set their width. We can use equation (3.6) to generate a set of curves which superficially resemble XANES spectra. Accordingly, the code for generating these spectra might look like (where we use the Numpy [61] and Matplotlib [62] software packages):

```
import numpy as np
import matplotlib.pyplot as plt
import random as random

X = np.linspace(-2,10,100) # Generate an evenly-spaced domain
N = 1000 # Number of fake spectra to generate
all_spectra = [] # Set up list of spectra to populate

A = np.random.uniform(0,1,N)
B = np.random.uniform(-.5,.5,N)
C = np.random.uniform(1,5,N)
D = np.random.uniform(1,5,N)
E = np.random.uniform(4,8,N)
F = np.random.uniform(1,2,N)
```

```
spectra = []
for n in range(10000):
    Y = toy_spectra(X,A[i],B[i],C[i],D[i],E[i],F[i])
    spectra.append(Y)

for y in spectra[:10]:
    plt.plot(X,y)
```

10 sample "spectra" are plotted below in Fig. 3.5.

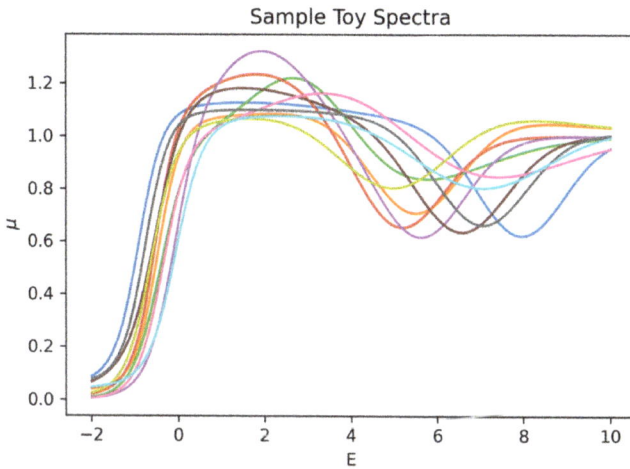

Fig. 3.5: 10 sample spectra generated with randomly chosen generating variables that exhibit clear variation in edge location, sharpness, and peak/trough location/size.

We can see that different peaks will be associated with different features. Now, we will demonstrate two different ways of extracting features associated with each spectrum using random forest models that differ in their input features. Standard packages make the training of random forest models easy; for our implementation, we will use and assume the syntax of Scikit-Learn [63]. To begin, we can fit a random forest to the appropriate data:

```
from sklearn.regressor import RandomForestRegressor
rf = RandomForestRegressor()
rf.fit(X,Y)
```

So here, the *features* which are input to the model are the individual values of absorption $\mu(X)$ on the x-axis. We use the variable X here to refer to the function values $\mu(X)$ of individual spectra and Y to refer to a particular parameter used to generate the corresponding spectrum (one of $\{a, b, c, d, e, f\}$). The model will work based on finding which features best "split" the training dataset in order to make predictions about respective regression values. The model "rf" will thus attempt to fit the spectral values X to the properties Y which generated the spectrum. Parity plots to models fit to each of the generating variables are included in Fig. 3.6.

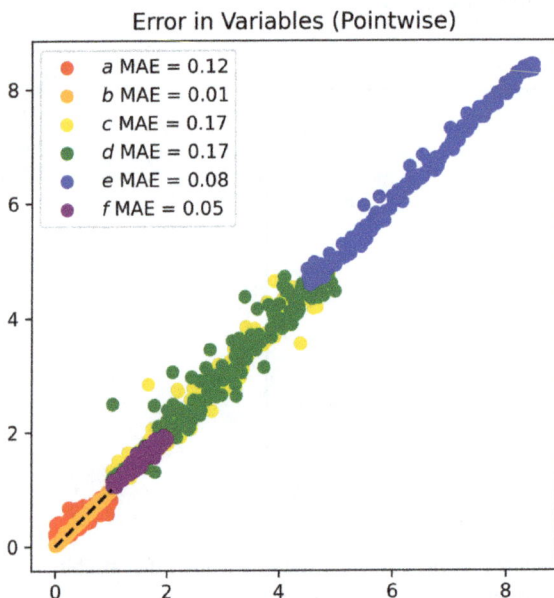

Fig. 3.6: Parity plot visualizing performance of the regression in recovering each of the generating variables for each of a, b, c, d, e, f. True values are along the x-axis, and recovered values are along the y. The departure from the central dotted line $y = x$ helps make it clear where systematic errors may be arising; for instance, over- or underestimation in certain regions of parameter space.

We can start with a simple parameter, a; the parameter which controls the sharpness of the curve. This parameter will be most consequential for the shape of the curve around $E = 0$, at which point, it begins to appreciably increase the argument in the exponent.

We can tell by inspection of equation (3.6) that the information about the parameter is "encoded" in the region where X is around [−1,1]. Can our model tell us that this is the region of importance?

First, we can fit a model to regress the generating variable a. Feature importance ranking of random forests and decision trees can help us to this end. The feature ranking values themselves are associated with reductions in the loss function used to train the model. We can visualize the relative importance of different features in different ways. One way is by normalizing the importances to the maximum score and seeing how they vary across the domain. We may color the domain of the plot based on the importance of each part of the feature space. In Scikit-learn, this is set using the "feature_ranking_" attribute of a random forest model.

```
for i, feature in enumerate(rf.feature_ranking_):
    plt.axvline(X[i],color=feature)
plt.show()
```

We can plot the feature ranking associated with each variable in Fig. 3.7, alongside 100 sample spectra, so it is easier to see which parts of the domain correspond to which parameter.

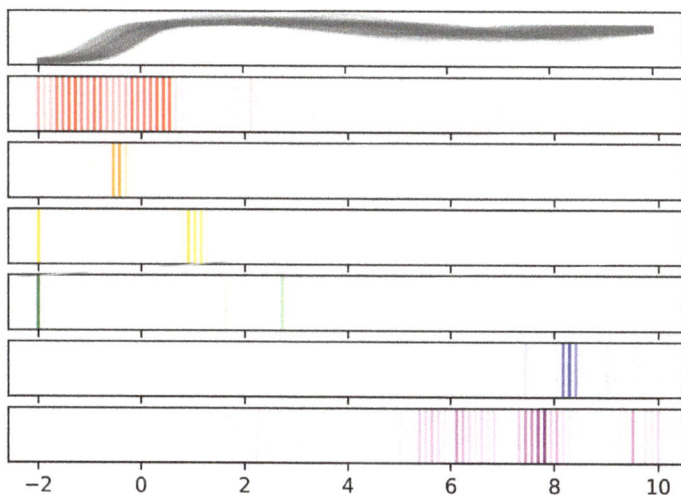

Fig. 3.7: Top: 100 sample spectra to help demonstrate which parts of the domain correspond to which parts of the spectrum. Bottom five plots: Feature ranking colored throughout the domain, where more heavily shaded regions indicate greater importance of different parts of the feature space. Variation between individual points within the same region can be explained by the stochastic nature of random forest training due to randomized selection of which variables are used to train different parts of the tree, in some iterations a region which is "truly" associated with signal will be ranked as more, or less important.

Now, this shows us that just as we expected, the domain where the information is "encoded" is highlighted: because a determines properties of the spectrum around the "edge" region, that part of the domain is highlighted as important to the regression task.

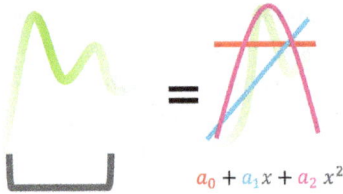

Fig. 3.8: A polynomial fit to a part of the spectrum 'encodes' the variation in the spectrum as a series of polynomial coefficients.

$$a_0 + a_1 x + a_2 x^2$$

Visually, we know that the size of the peak in said region is a reliable visual indicator of the magnitude of a. However, what if we did not know *how* a given part of the spectrum was contributing to the prediction task? Furthermore, is it possible that local correlations between points could be useful as well as the values of the spectrum at individual points? This behooves us to try an alternate means of featurization; we will continue by demonstrating how featurizing the spectra into polynomials can improve the performance, as well as help to shed light on how local correlations between individual spectral components contribute. This was the central contribution of Torrisi *et al.* [20].

We can perform such a featurization by breaking down the spectrum into individual subdomains or "chunks" across the energy values E, with code shown below:

```python
def polynomialize_by_idx(X:np.ndarray, Y:np.ndarray, N:int, deg:int=2)
                                    -> List[np.polynomial.polynomial]:
    """
    :param X: X domain
    :param Y: Y domain
    :param N: Number of splits over domain to make
    :param deg: degree of polynomial to use
    :return:
    """
    n_x = len(X)
    step = n_x // N

    domain_splits = list(range(0, n_x, step))
    if domain_splits[-1] != n_x - 1:
        domain_splits.append(n_x)

    # Perform fits
    polynomials = []
    for i in range(N):
        left = domain_splits[i]
        right = domain_splits[i + 1]
        x = X[left:right]
```

```
    y = Y[left:right]
    cur_poly, full_data = np.polynomial.Polynomial.fit(x, y, deg=deg,
    full=True)
    cur_poly.labels = [ f'$a_{n}:[{x[0]:.1f},{x[-1]:.1f}]$' for n in range
    (deg+1)]
    polynomials.append(cur_poly)
  return polynomials
```

Now, if we perform a random forest fit on the *coefficient* values instead of the values of the spectra themselves, we can obtain the spread of performances seen in Fig. 3.9.

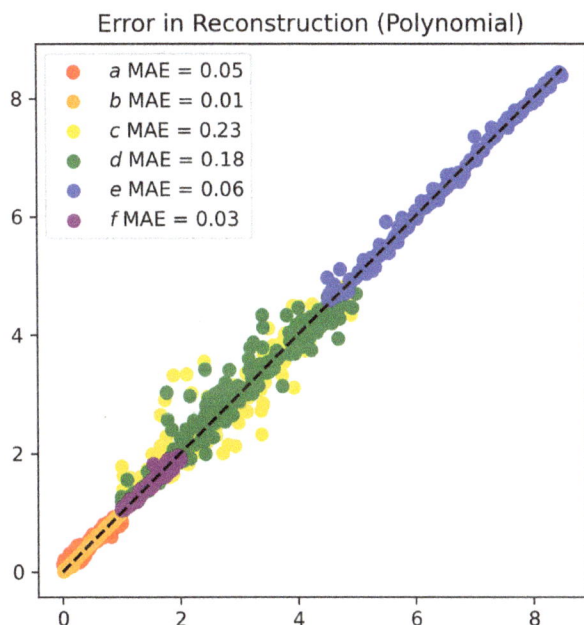

Fig. 3.9: Performance of models using polynomial featurization based on splitting the spectrum into 10 equal pieces and fitting quadratic polynomials to each part. The performance notably improves for prediction of *a*.

Now, when we look at the feature ranking associated with each individual feature, we see that it is much easier to digest what part of the spectrum contributes the most to the regression decision. For instance, in Fig. 3.10, note that the linear and quadratic terms derivative of the edge part of the spectrum are highlighted as the most important to the decision. Given that *a* determines the sharpness of the edge in equation (3.6), this demonstrates that the model is highlighting *a* in a way that provides a hint as to how it relates to the generating function of the spectra.

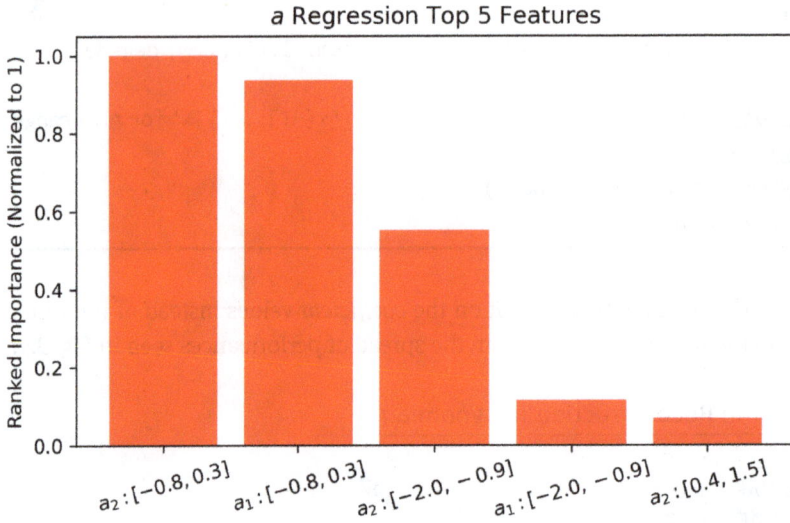

Fig. 3.10: The feature ranking associated with the regression of the a term from the spectra. The a_1 and a_2 terms, which roughly correspond to the linear/quadratic (first derivative/second derivative) trends of the edge-associated energy range are the most important for the eventual prediction of the generating parameter a.

We pause here to note that in evaluating the performance of the model, and the utility of this method of generating insights, we should do well to avoid the curse of knowledge – forgetting what it is like to not know. We clearly see the functional form that generated the function, but if we did not know *a priori* what the relationship between a and the spectra was, this method would help to tease it out moreso than a random forest acting on the unprocessed spectra would.

3.3.3.3 Remarks on extensibility

We note that if the reader is interested in applying this methodology to some other kind of spectral data, there is a lot of freedom in how featurization can be performed depending on the problem of interest. For data which rely on peak-like behavior, like XRD for single- or polycrystalline samples, the featurization could come from convoluting Gaussian peaks with the spectrum and using the value of the overlap as input spectra. For data which are represented as a sum of periodic signals, the use of a power spectrum from a Fourier transform may be more appropriate.

Note that the "fake" spectra which are generated in this way already have many helpful properties which set them apart from experimental spectra. In processing real experimental spectra, several checks must be ensured:

1. They will all lie along the same X-axis, with the same grid spacing and offset. This means that the feature vector of individual absorption values will correspond to the same energy grid for each data point. Many numerical packages (e.g. scipy's "Interp1d" function) are appropriate for this kind of procedure.
2. The data points have a well-defined label. Real experimental data which are used for training may require labels which are obtained from some other measurement, which may introduce its own uncertainty; for instance, coordination number of a given element in a material may be extracted from knowledge of the phase of the material, which may come from a different experimental measurement.
3. There is no measurement noise. In practical applications, these can be worked around in various ways; a smoothing filter could be applied in order to combat noise at each energy value, or the model can be trained on "gross" features which help to mitigate the sensitivity to noise in certain parts of the spectra.
4. We are not attempting to match theory with experiment. In real applications, theoretically generated XAS spectra may suffer from qualitative or quantitative errors owing to the level of theory which generated them; models which are trained on theoretical spectra which are transferred to experimental spectra must be approached with great care.

3.4 Future directions

We anticipate many benefits from the application of machine learning and AI to spectroscopy in the coming years. We anticipate spectroscopic methods to make a significant impact in several areas including experiment planning, analysis, design of instruments, and more. We discuss a few key approaches through which AI may enable these advances below:

3.4.1 The "forward problem" and surrogate modeling

Classes of "forward models" can in fact be very useful in the solution of an "inverse problem"; by recapitulating the input parameters to a signal generating function that encodes a materials' structure, the character of said structure can be determined once agreement has been achieved. This framework is entirely general, and can apply to many types of signals, including XAS. Essentially, one trains a "surrogate model" to replace an otherwise expensive calculation, such that it can be queried much more efficiently than that of the "oracle" which is the source of ground truth on which the model was trained. Also presented in the literature are so-called digital twins, which are *in silico* replacements of actual experiments.

As an interpolative tool, machine learning surrogates are useful for two primary reasons. First, once trained, inference will likely be orders of magnitude faster than the original process of generating the data. This is a huge advantage when the training data are expensive to compute, such as in real-world experiments, or in the case of computations that take a long time or require large amount of computing power. Second, they can be systematically improved in *e.g.* active learning frameworks. For example, suppose to analyze some hypothesis, it might take thousands of experiments, each of which is time consuming and monetarily expensive. Each experiment has some set of parameters, which perhaps represent different experimental conditions, and the relationship between those parameters and the result of each experiment, say a spectrum or scalar value, is unknown. One can imagine that instead of running thousands of experiments, it might only take 100 or fewer to train a model with a built-in measure of uncertainty like a Gaussian Process, such that the expected uncertainty for all possible experiments is reasonably low. In this way, the digital twin may be treated as a surrogate model for performing the experiment itself.

Forward modeling is an active research area in a large number of scientific domains. For example, forward modeling using Gaussian processes has been successfully shown to assist in optimizing properties of additively manufactured components [64]. In a completely different field, an "artificial chemist" has been constructed out of neural network ensembles to predict and tune the band gaps of inorganic perovskite quantum dots [65]. The applications of forward modeling are only growing, and will almost certainly continue to contribute to accelerating research and inference.

3.4.2 AI-assisted analysis of experiments

AI methods typically focus on automated analysis of spectroscopy data. However, several problems related to analysis of spectroscopy data such as assignment of peaks in an XPS spectrum, or fitting an EXAFS spectrum, are under-constrained problems making it difficult to identify a physically meaningful model from a range of mathematically equivalent models. One approach is to encode constraints into the machine learning model to eliminate extremely unphysical results. Even in such cases, a range of solutions are likely to be available for an expert to pick the best solution. By intertwining an expert decision-making process along with the results obtained from AI, one can develop a more complex decision-making agent that eventually learns to encode some of the expert decisions into the model constraints.

In addition, interpretable methods may also help to discover spectrum-property relationships that were not previously known [66] providing an accelerated knowledge feed to a human.

Machine learning methods could assist a human by providing an estimate and confidence intervals for the experimentally expensive measurements given that information from weakly related but easier to perform experiments are precollected. For example, in XAS measurements, EXAFS datasets require significantly higher measurement time to achieve reasonable signal to noise ratio compared to XANES. However, there are strong correlations between XANES and EXAFS datasets, so it may be possible to develop machine learning methods that predict EXAFS spectra given a XANES spectrum, thus providing a preview of EXAFS spectra even before a human makes a decision about whether it is worth spending experimental resources on collecting them. Quantified uncertainty on this prediction could help to assist in the eventual decision on whether to perform the follow-up measurement or not.

3.4.3 Automated planning of experiments

The more measurements that can be taken with limited beamtime, the more efficient the scientific process becomes. AI has already seen application in increasing the efficient use of beamtime at Lawrence Berkeley National Laboratory and the Institut Laue-Langevin in France. Autonomous planning of experiments are already underway using the gpCAM framework, based on Gaussian Process models [67] to optimize the rate of data collection within an experiment. Decisions about measurements are made automatically, conditioned on the results of previous measurements. A surrogate model trained in parallel with the measurements makes predictions about next steps based on the cost of a measurement as well as the expected "gain" in information. For further reading on experiments which seek to optimize a given parameter in a high-dimensional space, other chapters in this book describe the tradeoffs between "exploration" and "exploitation" in parameter space exploration. These algorithms are well suited toward the maximization or minimization of some objective function. We conjecture that an alternate framing could be set around scored certainty associated with qualitative classification, i.e. alternately minimizing the uncertainty of a prediction based on different available experiments. For example, this could mean determining the character of bonding between two elements in a compound. The decisions made by an autonomous algorithm or system could be constrained by any number of parameters, such as availability of data in literature or automatically queried online databases, availability of standards for calibration, etc.

More time-consuming or resource intensive modes of spectroscopy may be complemented by one another when assisted by machine learning. It is not uncommon for studies to involve the use of multiple imaging techniques targeted at different materials properties or functionalities. If machine learning can help reveal information encoded in different spectroscopic modalities that correlate with properties of interest, then the pace of experimental studies may be improved.

There may be opportunities for the discovery of new spectrum-property relationships that go beyond traditionally studied properties. For example, the topological classification of condensed matter systems is an exciting and relatively new field of study, and the characterization of topological materials is ordinary done with techniques such as angle-resolved photoemission spectroscopy (ARPES); Andrejevic *et al.* showed that the topological class of materials can be recovered via artificial neural networks and XANES spectra [66]. In this way, one modality of spectroscopy may be used to complement, or substitute for, another.

Nanoparticle size distributions and morphologies are difficult to resolve using XAS spectra alone, and typically rely on complementary modalities such as transmission electron microscopy; Timoshenko *et al.* have demonstrated how simulated XANES spectra can be used to train models which recapitulate nanoparticle size and morphology that are extensible to experimental data [68]. Similar to the example of topological systems, AI methods may help to allow modalities to complement or replace one another.

3.4.4 Generalized multimodal models

Some kind of generalized prediction algorithms which can combine inputs from multiple different modalities of spectroscopy could allow for more complicated inverse problems to be solved. For example, consider a tool which could combine input data from Raman, XAS, and XRD spectroscopy to determine structural and functional properties with high accuracy. The use of different knowledge sources, when combined with interpretable models, may help to detect subtle correlations across many data points which may be difficult to detect by individual domain experts, particularly when the output prediction space is very high dimensional.

Wherever possible, incorporating physics-based insight into these models or their internal representations of input data would help to reduce their data hunger and improve their utility. Consider that different forms of imaging on the same systems may result in correlated or redundant information; degrees of freedom associated with ligand oxygen bonds in a transition metal oxide may be sampled by both Raman spectra as well as XRD. A particular phase change may be difficult to resolve using Raman spectra, but easy to detect using XRD. The use of hidden physics-based variables in the model may help to avoid the use of large datasets. Physics insight to constrain model form and function should be used wherever possible. For a particular characterization question, it would be helpful to be able to plan out the minimum number of experiments and measurements that would be necessary to characterize the structural or functional properties.

However, a severely limiting factor to the development of these models is that generating experimental data may be costly and time consuming for a wide class of materials. The adoption of common standards and application programming interfaces (APIs) for the materials science community [69] may help experimentalists to share data with one another which could then be used to guide the development of such models. Tools which allow for the easy conversion of data into common formats as it is collected would be a boon to the materials spectroscopy community.

3.4.5 Algorithmic and experimental development

For sufficiently complicated multicomponent systems, existing theoretical approaches to generating XAS spectra may not be able to recapitulate the underlying spectrum with quantitative or qualitative accuracy, even when more sophisticated methods such as those which treat the excited electron state via solution of the Bethe–Salpeter equations. We anticipate future algorithmic developments on at least two fronts: Machine learning that can work exclusively from spectra in order to bypass the use of intermediate simulation methods, and algorithms which can learn corrections from high-quality computed spectra to match with experiment.

Additionally, new types of X-ray spectroscopy may become more accessible. For example, recent X-ray free electron laser sources like LCLS (the LINAC coherent light source) [70] provide broadband X-rays in a self-amplified spontaneous emission (SASE) mode. SASE produces intense light, but is stochastic, with spiky spectra in which its energy distribution and intensity change drastically from shot to shot within a band envelope. Traditionally, a monochromatized X-ray is used for absorption spectroscopy, by scanning through the absorption energy region of metals of interest. However, a SASE beam can be used for absorption spectroscopy without monochromatizing the X-ray beam: If the SASE spectral amplitude is accurately correlated with outgoing signals from the sample shot-by-shot, one can recover various resonant signals like X-ray absorption and Resonant Inelastic X-ray scattering (RIXS) [71].

References

[1]   Hanson H, Beeman WW. The Mn K absorption edge in manganese metal and manganese compounds. Phys Rev, Jul 1949, 76, 1, 118–121, doi: 10.1103/PhysRev.76.118.

[2]   Gregoire JM, Van Campen DG, Miller CE, Jones RJR, Suram SK, Mehta A. High-throughput synchrotron X-ray diffraction for combinatorial phase mapping. J Synchrotron Radiat, 2014, 21, 6, 1262–1268, doi: 10.1107/S1600577514016488.

[3]   Abuín M, Serrano A, Chaboy J, García MA, Carmona N. XAS study of Mn, Fe and Cu as indicators of historical glass decay. J Anal At Spectrom, 2013, 28, 7, 1118–1124, doi: 10.1039/c3ja30374h.

[4]   Bianchini M, et al. The interplay between thermodynamics and kinetics in the solid-state synthesis of layered oxides. Nat Mater, Oct 2020, 19, 10, 1088–1095, doi: 10.1038/s41563-020-0688-6.

[5]   Herrera-Gomez A, Bravo-Sanchez M, Ceballos-Sanchez O, Vazquez-Lepe MO. Practical methods for background subtraction in photoemission spectra. Surface Interface Anal, Oct 2014, 46, 10–11, 897–905, doi: 10.1002/sia.5453.

[6]   Routh PK, Liu Y, Marcella N, Kozinsky B, Frenkel AI. Latent Representation Learning for Structural Characterization of Catalysts. Am Chem Soc, Mar 2021, 2086–2094, doi: 10.1021/acs.jpclett.0c03792.

[7]   Chatzidakis M, Botton GA. Towards calibration-invariant spectroscopy using deep learning. Sci Rep, Dec 2019, 9, 1, 1–10, doi: 10.1038/s41598-019-38482-1.

[8]   Carbone MR, Yoo S, Topsakal M, Lu D. Classification of local chemical environments from X-ray absorption spectra using supervised machine learning. Phys Rev Mater, 2019, 3, 3, 33604, doi: 10.1103/PhysRevMaterials.3.033604.

[9]   Zheng C, Chen C, Chen Y, Ong SP. Random forest models for accurate identification of coordination environments from x-ray absorption near-edge structure, Patterns, 100013, Apr. 2020, doi: 10.1016/j.patter.2020.100013.

[10] Evans J. X-ray Absorption Spectroscopy for the Chemical and Materials Sciences, John Wiley & Sons, Inc., 2018.

[11] Tromp M, Moulin J, Reid G, Evans J. Cr K-edge XANES spectroscopy: Ligand and oxidation state dependence – What is oxidation state? AIP Conf Proc, 2007, 882, 699–701, doi: 10.1063/1.2644637.

[12] Yildirim B, Riesen H. Coordination and oxidation state analysis of cobalt in nanocrystalline LiGa5O8 by X-ray absorption spectroscopy. J Phys Conf Ser, 2013, 430, 1, 012011, doi: 10.1088/1742-6596/430/1/012011.

[13] Sarangi R, et al. Sulfur K-edge X-ray absorption spectroscopy as a probe of ligand-metal bond covalency: Metal vs ligand oxidation in copper and nickel dithiolene complexes. J Am Chem Soc, Feb 2007, 129, 8, 2316–2326, doi: 10.1021/ja0665949.

[14] Wong J, Lytle FW, Messmer RP, Maylotte DH. K-edge absorption spectra of selected vanadium compounds. Phys Rev B, Nov 1984, 30, 10, 5596–5610, doi: 10.1103/PhysRevB.30.5596.

[15] Mueller DN, MacHala ML, Bluhm H, Chueh WC. Redox activity of surface oxygen anions in oxygen-deficient perovskite oxides during electrochemical reactions. Nat Commun, 2015, 6, doi: 10.1038/ncomms7097.

[16] Farges F, Brown GE, Navrotsky A, Gan H, Rehr JJ. Coordination chemistry of Ti(IV) in silicate glasses and melts: II. Glasses at ambient temperature and pressure. Geochim Cosmochim Acta, 1996, 60, 16, 3039–3053, doi: 10.1016/0016-7037(96)00145-7.

[17] Farges F, Brown GE. Ti-edge XANES studies of Ti coordination and disorder in oxide compounds: Comparison between theory and experiment. Phys Rev B – Condensed Mat Mater Phys, 1997, 56, 4, 1809–1819, doi: 10.1103/PhysRevB.56.1809.

[18] Farges F, Brown GE, Petit PE, Munoz M. Transition elements in water-bearing silicate glasses/ melts. Part I. A high-resolution and ancharmonic analysis of Ni coordination environments in crystals,glasses,and melts. Geochim Cosmochim Acta, 2001, 65, 10, 1665–1678, doi: 10.1016/S0016-7037(00)00625-6.

[19] Jackson WE, et al. Multi-spectroscopic study of Fe(II) in silicate glasses: Implications for the coordination environment of Fe(II) in silicate melts. Geochim Cosmochim Acta, Sep 2005, 69, 17, 4315–4332, doi: 10.1016/j.gca.2005.01.008.

[20] Torrisi SB, et al. Random forest machine learning models for interpretable X-ray absorption near-edge structure spectrum-property relationships. Npj Comput Mater, 2020, 6, 1, doi: 10.1038/s41524-020-00376-6.

[21] Suntivich J, Gasteiger HA, Yabuuchi N, Nakanishi H, Goodenough JB, Shao-Horn Y. Design principles for oxygen-reduction activity on perovskite oxide catalysts for fuel cells and metal-air batteries. Nat Chem, Jun 2011, 3, 7, 546–550, doi: 10.1038/nchem.1069.

[22] Stern EA, Sayers DE, Lytle FW. Extended x-ray-absorption fine-structure technique. III. Determination of physical parameters. Phys Rev B, 1975, 11, 12, 4836–4846, doi: 10.1103/PhysRevB.11.4836.

[23] Stern EA. Theory of the extended x-ray-absorption fine structure. Phys Rev B, 1974, 10, 8, 3027–3037, doi: 10.1103/PhysRevB.10.3027.

[24] Sayers DE, Stern EA, Lytle FW. New technique for investigating noncrystalline structures: Fourier analysis of the extended x-ray-absorption fine structure. Phys Rev Lett, Nov 1971, 27, 18, 1204–1207, doi: 10.1103/PhysRevLett.27.1204.

[25] Rehr JJ, Albers RC. Theoretical approaches to x-ray absorption fine structure. Rev Mod Phys, 2000, 72, 3, 621–654, doi: 10.1103/RevModPhys.72.621.

[26] Norman P, Dreuw A. Simulating X-ray Spectroscopies and Calculating Core-Excited States of Molecules. Chem Rev, 2018, 118, 15, 7208–7248, doi: 10.1021/acs.chemrev.8b00156.

[27] Gallagher M, Deacon P. Neural networks and the classification of mineralogical samples using x-ray spectra, in ICONIP 2002 – proceedings of the 9th international conference on neural information processing: Computational intelligence for the e-age, 2002, vol. 5, pp. 2683–2687. doi: 10.1109/ICONIP.2002.1201983.

[28] Lüder J. Determining electronic properties from L -edge x-ray absorption spectra of transition metal compounds with artificial neural networks. Phys Rev B, Jan 2021, 103, 4, 045140, doi: 10.1103/PhysRevB.103.045140.

[29] Timoshenko J, Frenkel AI. "Inverting" X-ray absorption spectra of catalysts by machine learning in search for activity descriptors. ACS Catal, 2019, 9, 11, 10192–10211, doi: 10.1021/acscatal.9b03599.

[30] Ameh ES. A review of basic crystallography and x-ray diffraction applications. Int J Adv Manuf Technol, Dec 2019, 105, 7–8, 3289–3302, doi: 10.1007/s00170-019-04508-1.

[31] Lee JW, Park WB, Lee JH, Singh SP, Sohn KS. A deep-learning technique for phase identification in multiphase inorganic compounds using synthetic XRD powder patterns. Nat Commun, Dec 2020, 11, 1, 1–11, doi: 10.1038/s41467-019-13749-3.

[32] Ali A, et al. Machine learning accelerated recovery of the cubic structure in mixed-cation perovskite thin films. Chem Mater, Apr 2020, 32, 7, 2998–3006, doi: 10.1021/acs.chemmater.9b05342.

[33] Jones RR, Hooper DC, Zhang L, Wolverson D, Valev VK. Raman techniques: Fundamentals and frontiers, vol. 14, Springer New York LLC, Dec. 2019. doi: 10.1186/s11671-019-3039-2.

[34] Devereaux TP, Hackl R. Inelastic light scattering from correlated electrons. Rev Mod Phys, 2007, 79, 1, 175–233, doi: 10.1103/RevModPhys.79.175.

[35] Cui A, et al. Decoding phases of matter by machine-learning raman spectroscopy. Phys Rev Appl, Nov 2019, 12, 5, doi: 10.1103/PhysRevApplied.12.054049.

[36] Lee W, Lenferink ATM, Otto C, Offerhaus HL. Classifying Raman spectra of extracellular vesicles based on convolutional neural networks for prostate cancer detection. J Raman Spectr, Feb 2020, 51, 2, 293–300, doi: 10.1002/jrs.5770.

[37] Ament SE, et al. Multi-component background learning automates signal detection for spectroscopic data. Npj Comput Mater, Dec 2019, 5, 1, 77, doi: 10.1038/s41524-019-0213-0.

[38] Sonneveld EJ, Visser JW. Automatic collection of powder data from photographs. J Appl Crystallogr, 1975, 8, 1–7, Available: http://scripts.iucr.org/cgi-bin/paper?a12580.

[39] Tougaard S. Algorithm for automatic X-ray photoelectron spectroscopy data processing and X-ray photoelectron spectroscopy imaging. J Vacuum Sci Technol A: Vacuum, Surfaces, Films, 2005, 23, 4, 741–745, doi: 10.1116/1.1864053.

[40] Alberi K, et al. The 2019 materials by design roadmap. J Phys D: Appl Phys, Jan 2019, 52, 1, 48, doi: 10.1088/1361-6463/aad926.

[41] Seah MP. The quantitative analysis of surfaces by XPS: A review. Surface Interface Anal, 1980, 2, 6, 222–239, doi: 10.1002/sia.740020607.

[42] Laue M. Über die Interferenzerscheinungen an planparallelen Platten. Ann Phys, 1904, 318, 1, 163–181, doi: 10.1002/andp.18943180107.

[43] Palmer EM, Horowitz TS, Torralba A, Wolfe JM. What are the shapes of response time distributions in visual search?. J Exp Psychol Hum Percept Perform, 2011, 37, 1 58–71.

[44] Golubev A. Exponentially modified gaussian (EMG) relevance to distributions related to cell proliferation and differentiation. J Theor Biol, 2010, 262, 2 257–266.

[45] Hattrick-Simpers JR, Gregoire JM, Kusne AG. Perspective: Composition–structure–property mapping in high-throughput experiments: Turning data into knowledge. APL Mater, 2016, 4, 5 053211.

[46] Long CJ, Bunker D, Li X, Karen VL, Takeuchi I. Rapid identification of structural phases in combinatorial thin-film libraries using x-ray diffraction and non-negative matrix factorization. Rev Sci Instrum, 2009, 80, 103902, doi: 10.1063/1.3216809.

[47] Stanev V, Vesselinov VV, Kusne AG, Antoszewski G, Takeuchi I, Alexandrov BS. Unsupervised phase mapping of X-ray diffraction data by nonnegative matrix factorization integrated with custom clustering. Npj Comput Mater, 2018, 4, 1 43.

[48] Xue Y et al., Phase-Mapper: an AI platform to accelerate high throughput materials discovery, 2017, https://arxiv.org/abs/1610.00689.

[49] Suram SK, et al. Automated phase mapping with AgileFD and its application to light absorber discovery in the V-Mn-Nb oxide system. ACS Comb Sci, Jan 2017, 19, 1, 37–46, doi: 10.1021/acscombsci.6b00153.

[50] Gomes CP, et al. CRYSTAL: A multi-agent AI system for automated mapping of materials' crystal structures. MRS Commun, Jun 2019, 9, 2, 600–608, doi: 10.1557/mrc.2019.50.

[51] Long C, et al. Rapid structural mapping of ternary metallic alloy systems using the combinatorial approach and cluster analysis. Rev Scientific Inst, 2007, 78, 7, 072217.

[52] Kusne AG, Keller D, Anderson A, Zaban A, Takeuchi I. High-throughput determination of structural phase diagram and constituent phases using GRENDEL. Nanotechnology, 2015, 26, 44 444002.

[53] Iwasaki Y, Kusne AG, Takeuchi I. Comparison of dissimilarity measures for cluster analysis of X-ray diffraction data from combinatorial libraries. Npj Comput Mater, 2017, 3, 1 4.

[54] Kusne AG, et al. On-the-fly closed-loop materials discovery via Bayesian active learning. Nat Commun, Nov 2020, 11, 1, 5966, doi: 10.1038/s41467-020-19597-w.

[55] Rossi F, Van Beek P, Walsh T. Handbook of Constraint Programming, Elsevier, 2006.

[56] Elser V, Rankenburg I, Thibault P. Searching with iterated maps. Proceed Nat Acad Sci, Jan 2007, 104, 2, 418–423, doi: 10.1073/pnas.0606359104.

[57] Amizadeh S, Matusevych S, Weimer M. PDP: A general neural framework for learning constraint satisfaction solvers, arXiv preprint arXiv:1903.01969, 2019.

[58] Bai J, et al. Phase-mapper: Accelerating materials discovery with AI. AI Mag, Mar 2018, 39, 1, 15–26, doi: 10.1609/aimag.v39i1.2785.

[59] Lipton ZC. The mythos of model interpretability. Commun ACM, 2018, 61, 10, 35–43, doi: 10.1145/3233231.

[60] Breiman L. Random forests. Mach Learn, Oct 2001, 45, 1, 5–32, doi: 10.1023/A:1010933404324.

[61] Harris CR, et al. Array programming with NumPy. Nature, Sep 2020, 585, 7825, 357–362, doi: 10.1038/s41586-020-2649-2.

[62] Hunter JD. Matplotlib: A 2D graphics environment. Comput Sci Eng, 2007, 9, 3, 90–95, doi: 10.1109/MCSE.2007.55.

[63] Pedregosa F, et al. Scikit-learn: Machine learning in Python. J Mach Learn Res, 2011, 12, 2825–2830.

[64] Gongora AE, et al. A Bayesian experimental autonomous researcher for mechanical design. Sci Adv, Apr 2020, 6, 15, eaaz1708, doi: 10.1126/sciadv.aaz1708.

[65] Epps RW, et al. Artificial chemist: An autonomous quantum dot synthesis bot. Adv Mater, 2020, 32, 30, 2001626.

[66] Andrejevic N, Andrejevic J, Rycroft CH, Li M. Machine learning spectral indicators of topology, arXiv, Mar. 2020, http://arxiv.org/abs/2003.00994

[67] Noack MM, Yager KG, Fukuto M, Doerk GS, Li R, Sethian JA. A Kriging-Based Approach to Autonomous Experimentation with Applications to X-Ray Scattering. Sci Rep, Dec 2019, 9, 1, 1–19, doi: 10.1038/s41598-019-48114-3.

[68] Timoshenko J, Lu D, Lin Y, Frenkel AI. Supervised machine-learning-based determination of three-dimensional structure of metallic nanoparticles. J Phys Chem Lett, 2017, 8, 20, 5091–5098, doi: 10.1021/acs.jpclett.7b02364.

[69] Andersen, C.W., Armiento, R., Blokhin, E. et al. OPTIMADE, an API for exchanging materials data. Sci Data 8, 217 (2021). https://doi.org/10.1038/s41597-021-00974-z.

[70] Sierra RG, et al. The macromolecular femtosecond crystallography instrument at the linac coherent light source 1. J Synchrotron Rad, 2019, 26, 346–357, doi: 10.1107/S1600577519001577.

[71] Fuller, F., Gul, S., Chatterjee, R. et al. Drop-on-demand sample delivery for studying biocatalysts in action at X-ray free-electron lasers. Nat Methods 14, 443–449 (2017). https://doi.org/10.1038/nmeth.4195.

Benjamin P. MacLeod, Fraser G. L. Parlane, Amanda K. Brown,
Jason E. Hein, Curtis P. Berlinguette

4 Flexible automation for self-driving laboratories

4.1 Introduction

It often takes decades to commercialize a new material [1]. To help reduce the time it takes to bring new materials to market, the materials science community has recently started to develop *self-driving laboratories* [2, 3]. These laboratories combine automation with data-driven experiment planning algorithms to efficiently search large parameter spaces for new, high-performance materials [4–14].

A key challenge in developing self-driving laboratories is that many materials science experiments are difficult to automate. Materials scientists use a wide range of specialized experimental procedures and sample types. The apparatus used to perform these procedures is often custom built and one of a kind. Materials scientists are therefore not able to leverage standardized, commercially available automated tools to the extent that is possible in other disciplines, such as the life sciences.

Instead, materials researchers seeking to leverage the benefits of automation must often develop their own, highly customized automated experiments [15–17]. Deploying these tools requires expertise and resources which are unavailable to most laboratories. As a result, most materials scientists still perform their experiments by hand and self-driving laboratories are out of reach for most organizations.

The ongoing commoditization of robotics offers an opportunity to increase the accessibility of automated experiments. In the last decade, versatile robotic arms suitable for automating a wide range of materials science experiments have become much safer, less expensive, and more user-friendly [18]. Here we explore "flexible automation": the use of robotics to build automated experiments that can easily be reconfigured. We provide examples of robotic experiments and self-driving laboratories built using flexible automation, followed by a discussion of strategies and technologies that can be used in concert with flexible automation to build more effective self-driving laboratories.

Benjamin P. MacLeod, Fraser G. L. Parlane, Amanda K. Brown, Jason E. Hein,
Curtis P. Berlinguette, Department of Chemistry, University of British Columbia, Canada

https://doi.org/10.1515/9783110738087-004

4.2 Flexible automation for materials science workflows

The first industrial robotic arm – the *Unimate* – was deployed in 1961 to unload automotive parts from a die-casting machine in a General Motors factory [19]. The versatility of robots was quickly recognized and robotic arms were soon applied to a range of other repetitive manufacturing tasks, including welding and painting [19]. Despite this versatility, the size, cost, hazards, and complexity of robotic arms restricted their use to industrial settings for many decades.

The use of robotic arms for materials research has, until recently, occurred predominantly in large organizations (e.g., corporate labs [20] or user facilities [21]). These organizations have the resources to support expensive robotic systems requiring specialized programming languages and professional safety engineering. Even for large organizations, the cost and time required to modify such systems inhibits the automation of exploratory materials research requiring frequent modifications to the apparatus and procedures. This situation is beginning to change, however, thanks to *flexible automation*.

Flexible automation makes automation faster and easier to develop by leveraging safe, lower cost, and easy-to-use robotics. Robots with these characteristics only started to become available with the introduction of intrinsically safe 6-axis robotic arms by Universal Robotics in 2008 [22]. Also known as collaborative robots, these robots have collision force limits that protect nearby workers and equipment from harm and eliminate the need for costly and restrictive safety enclosures. The intrinsic safety of these robots, along with their plummeting prices, steadily improving capabilities, and ease of use are causing them to proliferate in a wide range of applications [18], including laboratory research [7–9, 13, 23–30].

These advances in robotics, along with improvements in open-source software, low-cost sensors and electronics, rapid prototyping, machine vision, and other technologies now provide researchers with a flexible automation toolkit. This toolkit makes it faster, easier, and more cost-effective to automate specialized materials science experiments. Flexible automation empowers researchers to rapidly develop and reconfigure automated materials science workflows.

An automated workflow is a sequence of operations carried out by a system without ongoing human intervention. An automated workflow for materials science typically involves steps for the synthesis, processing, or characterization of materials. The application of flexible automation to a generic automated materials science workflow is shown in Fig. 4.1. In such workflows, a robot typically transports samples between stationary hardware modules that perform synthesis, processing, or characterization tasks. The use of a robot that can pick up tools to perform tasks at multiple locations further expands the scope of operations possible with the system. The robot's actions and the operation of the hardware modules are controlled

Experimental workflow

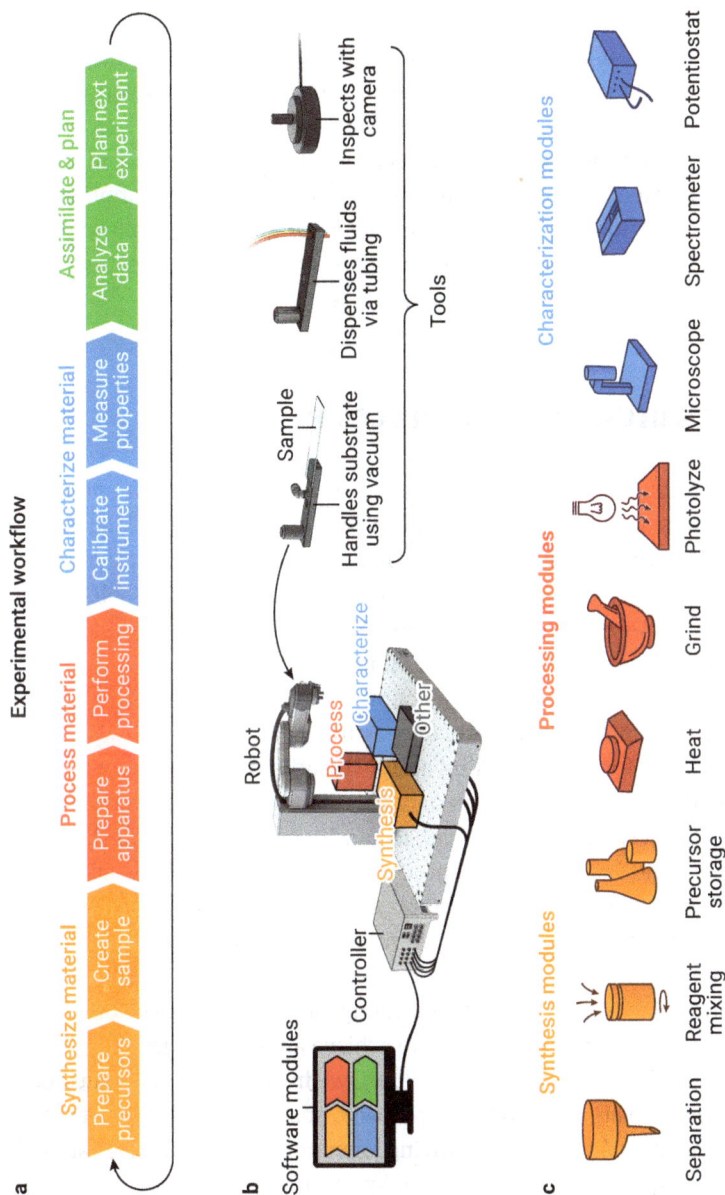

Fig. 4.1: Flexible automation involves equipping a robot with modules to enable an automated experimental workflow. First, the experimental workflow to be automated is broken down into a sequence of steps (a) A suitable flexible automation platform is then selected and equipped with the hardware modules, tools, and software modules necessary for the robot to perform each step in the workflow (b). Many types of hardware modules and tools required by the workflow can be used; some examples suitable for common experimental steps are shown (c).

by software modules through one or more controllers. By combining suitable robots and controllers with appropriate tools, hardware modules, and software modules, a wide range of customized automated experiments can be quickly developed. If research needs change, the capabilities of the flexible automation system can quickly be modified by adding or replacing modules as needed. Once the experimental steps are successfully automated, data analysis and algorithmic experiment planning steps can be added to the workflow to create a self-driving laboratory. In the following section, we provide examples of how flexible automation is enabling the construction of new kinds of self-driving laboratories for materials science.

4.3 Flexible automation in practice

We recently used flexible automation to build a self-driving laboratory for optimizing spin-coated thin films. Our first study using this laboratory optimized the charge carrier mobility of an organic hole transport material (HTM) used in perovskite solar cells [8] (Fig. 4.2; supported by Natural Resources Canada). The self-driving laboratory employed a 4-axis SCARA-type robot (N9, North Robotics) and a variety of hardware modules to prepare precursors, spincoat these precursors onto glass substrates, anneal the resulting films, and characterize the hole mobility by combining conductivity and optical absorbance measurements. The N9 robot features a built-in pipette mount for fluid handling and a versatile pneumatic gripper suitable for gripping cylindrical objects such as vials and vial caps. This gripper can also pick up appropriately shaped tools (e.g., tools with cylindrical handles). We developed a substrate-handling tool to transport the glass substrates required for our spin-coating workflow. The N9 robot also features a versatile controller which provides a variety of convenient hardware-software interfaces. These interfaces expedited the development of controls for the various hardware modules required in our film optimization workflow, including a spin-coater, an annealing module, and a spectrometer.

After fine-tuning the automated workflow to achieve an adequate level of reliability (e.g., 20–30 experimental cycles before a failure), we incorporated automated data analysis and experiment planning into the workflow to enable closed-loop experimentation. Our first demonstration of this self-driving laboratory [8] used a Bayesian optimization algorithm [31] to iteratively vary the doping level and annealing time to maximize the hole mobility of the HTM. An approximate global optimum for the hole mobility in this two-parameter material design space was identified in 30 experiments.

The flexible automation toolkit we employed to build our self-driving laboratory for thin films can readily be adapted to additional workflows using different hardware modules or tools. Fig. 4.3 shows examples of some of the diverse chemistry and materials science workflows we have tested in our labs using the same robot described above. In addition to spin coating, these workflows include adhesive

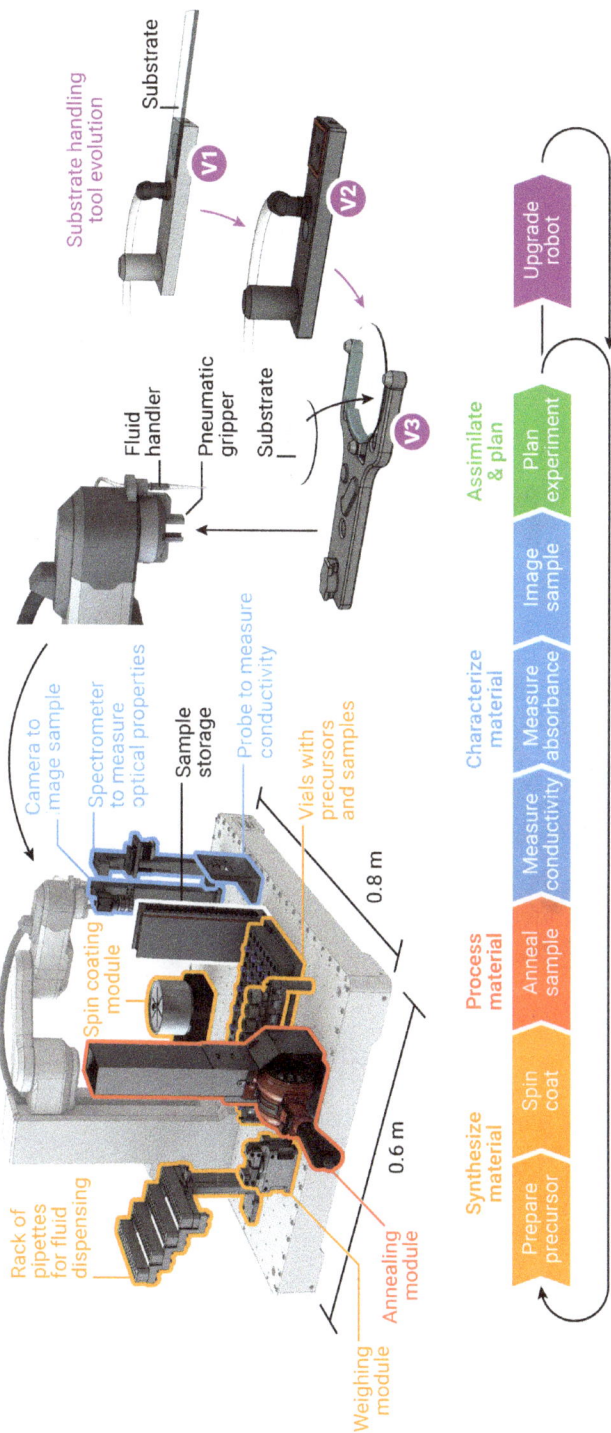

Fig. 4.2: A flexible automation platform configured for a thin-film optimization workflow. This workflow leveraged XYZ motion, fluid handling, and object gripping by the robot, employed a tool for substrate handling, and used hardware modules for film deposition, processing, and characterization. The steps in the workflow included gravimetric preparation of precursor solutions, film deposition by spin coating, annealing of the thin-film sample, and optical and electrical characterization. Automated data analysis and machine-learning-based experiment planning enabled autonomous experiments. Upgrades to the platform are made based on experience gained. For example, the figure shows the evolution of the substrate-handling tool from a vacuum-based all-metal design (V1) to a design featuring a silicone gasket to improve the vacuum seal reliability (V2) to a passive mechanical design not requiring vacuum (V3).

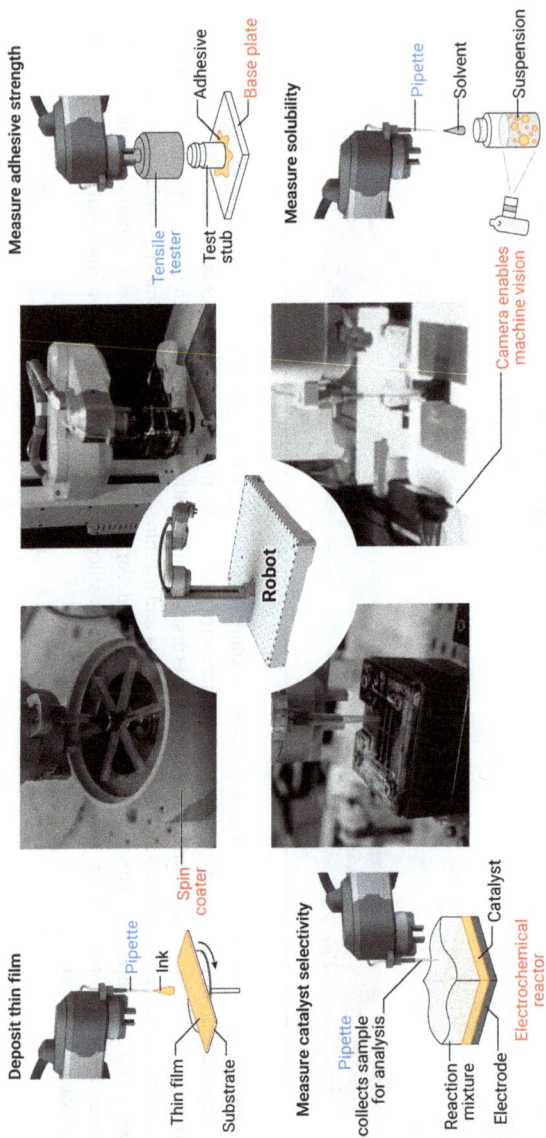

Fig. 4.3: Flexible automation enables diverse automated workflows. Several automated workflows in use in our laboratories are pictured. The same robot platform, with appropriately designed tools (blue) and modules (red), can fabricate thin films, measure the tensile strength of adhesives, evaluate the performance of heterogeneous catalysts, and quantify solubility. Photo credit: UBC.

strength testing (development supported by 3M), catalyst characterization (development supported by Suncor Energy Inc), and solubility screening [32] (development supported by Amgen Inc). These examples highlight how a flexible automation system can be repurposed for different applications. Another important benefit of flexible automation is the ability to upgrade an existing system to expand its capabilities.

To expand the capabilities of our original self-driving laboratory for thin films [8], we integrated our N9 robot with a larger, 6-axis collaborative robot arm (UR5e, Universal Robotics) [13]. The 850-nm reach of this arm enabled us to add larger hardware modules to our self-driving laboratory, including an X-ray fluorescence (XRF) microscope (M4 Tornado, Bruker) nearly equal in size to the N9 robot. To transfer samples to and from this XRF microscope, we equipped the UR5e robot with a substrate-handling tool similar to the one used by our N9 robot. The XRF microscope was readily equipped with a custom sample fixture suitable for automated loading and unloading of samples by the UR5 arm. The ability to measure the thickness of inorganic films using the XRF enabled our self-driving laboratory to perform a range of new workflows, including a multiobjective optimization of combustion-synthesized palladium films [13].

Other strategies for expanding flexible automation systems include placing robot arms on linear stages [28] or on mobile robotic vehicles [27]. These approaches have been used to develop some of the most sophisticated automated materials science workflows reported to date.

Brabec and coworkers have developed a flexible automation system called AMANDA capable of fabricating and testing entire photovoltaic devices [28, 29]. This system uses two SCARA-type collaborative robots on range-extending linear rails (PF400, Precise Automation) to link together an array of hardware modules for fabricating, processing, and characterizing thin films and multilayer devices inside a glovebox. These hardware modules include pipetting robots for precursor formulation and spin coating, a physical vapor deposition module for depositing charge selective and metallic contacts, thermal annealing stations, and a variety of optical and electronic characterization instruments. Special carriers enable batches of precursor vials or 25 mm x 25 mm glass substrates to be transported around the system. These carriers use a form factor that matches the SBS format, a widely used standard in biology automation [33]. Leveraging this standard is highly beneficial for flexible automation projects because numerous commercially available consumables (e.g., Microtiter™ plates), robotic grippers, and measurement instruments (e.g., plate readers) are compatible with this standard. This sophisticated flexible automation system can rapidly test fabrication conditions for organic photovoltaic cells and has already been used to fabricate devices with 14% power conversion efficiency [29]. The broad capabilities of this flexible automation system suggest it will be a valuable platform for research on photovoltaics and other classes of thin-film devices.

Materials that harvest solar energy were also the initial research target of Cooper and coworkers' *mobile robotic chemist*. This flexible automation system employed a collaborative robot arm (LBR iiwa, Kuka AG) mounted to a mobile robotic vehicle (Kuka Mobile Platform, Kuka AG) to transport photocatalyst samples between benchtop hardware modules distributed throughout an entire 70-m^2 laboratory. This system has the benefit of nearly unlimited working volume. This benefit comes at the cost of the additional complexity involved with navigating the mobile robot to each benchtop station and then accurately aligning the robotic arm to the station. The robotic arm used in this system has a multipurpose gripper designed to grasp single vials, racks of 16 vials, and solid-dispensing cartridges. This system can perform an automated workflow involving formulating complex photocatalyst mixtures by combining both powders and liquids in vials, inerting and sealing these vials, irradiating the photocatalysts, and subsequently quantifying the reaction products by gas chromatography. By combining this workflow with a Bayesian optimization algorithm, a photocatalyst mixture containing up to nine components was autonomously optimized over more than 600 experiments.

The above examples illustrate how flexible automation is applicable to a wide range of materials science experiments. While these experiments can range in scale from benchtop systems to entire laboratories, they all leverage versatile robotic arms to implement automated materials science workflows for which automation solutions are not commercially available. A survey of the literature containing additional examples of self-driving labs employing flexible automation is given in Tab. 4.1. In the following sections, we discuss additional technologies and strategies that help make flexible automated experiments faster to deploy and more effective.

Tab. 4.1: Examples of self-driving labs employing flexible automation in the literature.

Reference	Synopsis of workflow	Robot(s)	Synthesis and processing modules	Characterization modules	Tools
Gongora et al. 2020 [7]	Maximize compressive strength of a 3D-printed structure by varying four geometric design parameters	6-axis articulated robotic arm	3D printers (x6)	Compressive strength tester Analytical balance	–
MacLeod et al. 2020 [8]	Maximize hole mobility of an organic semiconductor by varying doping and annealing	4-axis SCARA robotic arm	Fluid dispensing Fluid mixing Spin coating Annealing	UV-VIS spectroscopy Conductance Optical imaging	Substrate handler

Tab. 4.1 (continued)

Reference	Synopsis of workflow	Robot(s)	Synthesis and processing modules	Characterization modules	Tools
Lochmüller and Lung, 1986 [24]	Maximize the absorbance response of a spectrophotometric protocol for phosphate quantification by varying concentrations of two reagents	3-axis cylindrical robotic arm	Fluid dispensing Fluid mixing Fluid heating	UV-VIS spectroscopy	Gripper tool Syringe tool
Burger et al. [27]	Maximize hydrogen evolution rate of a photocatalyst formulation by varying relative amounts of 10 different components	6-axis articulated robot arm mounted to a mobile robotic vehicle	Fluid dispensing Powder dispensing Mixing by sonication Capping/ inertization Photolysis	Gas chromatograph	Vial carrier rack
MacLeod et al. 2021 [13]	Identify optimal tradeoffs between processing temperature and conductivity for a solution processed metallic coating by varying temperature and relative amounts of three precursors	4-axis SCARA robotic arm 6-axis articulated robotic arm	Fluid dispensing Fluid mixing Drop casting Annealing	Optical imaging XRF microscopy Conductance	Substrate handler
Wagner et al. 2021 [28, 29]	Maximize power conversion efficiency of an organic solar cell by screening 10 device composition and processing parameters	4-axis SCARA robotic arm with range extender (x2) Cartesian pipetting robot (x2) 6-axis articulated robotic arm	Fluid dispensing Fluid mixing Spin coating Annealing Physical vapor deposition	UV-VIS spectroscopy Optical imaging Photovoltaic cell JV curve measurement	Substrate carrier tray Vial carrier rack

Tab. 4.1 (continued)

Reference	Synopsis of workflow	Robot(s)	Synthesis and processing modules	Characterization modules	Tools
Li et al. 2020 [9]	Maximize optical chirality of a nanoparticle by varying synthesis temperature and reagent concentration	6-axis articulated robotic arm	Microfluidic synthesis reactor	UV-VIS spectroscopy Circular dichroism spectroscopy	–

4.4 Rapid prototyping

Rapid prototyping is a key reason flexible automation platforms can be adapted to different workflows so quickly. Prototyping once required the creation of drawings that were sent to a machine shop for manufacture, with some delay before parts were returned for testing. Any design refinements would require the process to be repeated, and obtaining useful robotic hardware would often take weeks or months. Our design-build-test cycles now take hours or days. This is possible because of advances in 3D printing and CNC cutting tools, and because we can do virtual testing of new robotics components prior to builds. We use a 3D simulator driven by the same software commands as the physical robot to virtually test robot motions, robot/module interactions, and control code. Performing this virtual testing before the build saves time and money.

We used rapid prototyping to, for example, develop a tool that could enable the robot to handle glass substrates. The pneumatic gripper of our robot was not suitable for this task, but the new tool enabled the robot to transport the substrates without ever contacting (and contaminating) the surface containing the sample of interest. The tool also underwent many design iterations that would have not been considered without actually performing the experiments (see slide handler evolution in Fig. 4.2). We had similar experiences in quickly adding annealing modules (Fig. 4.4), chemical reactors, conductivity probes, cameras, and spectrometers to our automated platforms. These examples illustrate how rapid prototyping and the adaptability of a flexible automation platform enable capabilities to be added to automated workflows in a matter of days rather than months.

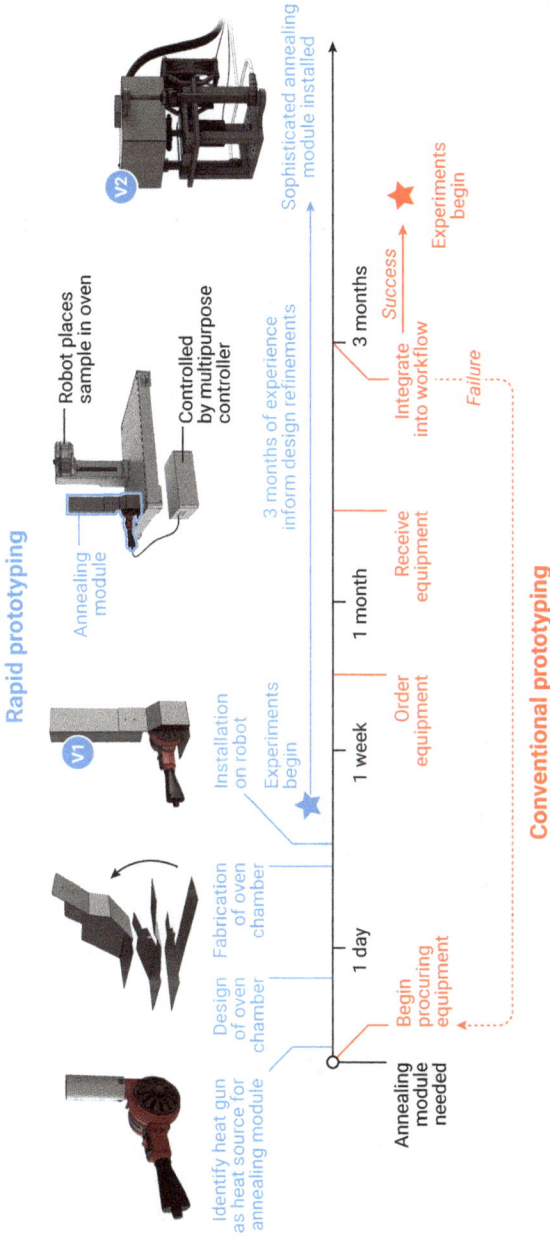

Fig. 4.4: Rapid prototyping of a thermal annealing module. Upon identifying the experimental need for an annealing module, a standard heat gun was repurposed for this application. An oven chamber for attachment to the heat gun was then designed from sheet metal and standard tubing suitable for rapid fabrication using a waterjet cutter. The interaction of the robot and sample with the annealer geometry was tested virtually using a robot simulator. The annealer was then mounted on the robot and on–off heating control was quickly implemented using a relay-controlled power port on the controller. The annealer was integrated into the automated experimental workflow using code already written for the virtual testing, and experiments involving annealing began. Knowledge gained over several months of experiments with the original annealer (V1) was used to develop a refined design (V2). The timeline shown for the rapid prototyping of the annealing module is the actual development timeline that occurred in our laboratory. For comparison, a representative timeline for procuring a commercial annealing module (assuming a standard design was commercially available) is shown.

4.5 The need for fast measurements

Automated experiments are most effective when large amounts of highly informative data are rapidly produced. It is therefore important to identify and address bottlenecks when developing an automated workflow.

While parallelized experimental methods [15–17, 34–36] increase throughput and should be leveraged where possible, these methods can be difficult to implement and are not applicable to every type of experiment. For applications where straightforward parallelized approaches are not available, automation is often not used at all. This need not be the case. Automating serial experiments can still increase throughput while providing many other benefits, including better data management and higher reproducibility. Moreover, serial experiments are often simpler to automate than parallel experiments and are therefore more easily integrated into an automated workflow.

Because materials synthesis and processing are generally easier to parallelize than characterization, the most challenging aspect of automated workflow design is often to identify sufficiently fast measurements. Consider our HTM case study where we needed to determine the hole mobility of each sample. This measurement normally requires the construction of multi-layered devices, [37] a procedure too complex and time-intensive for our automated workflow. We therefore invented a surrogate measurement that reports on the *relative* hole mobilities for a series of HTMs. Using an optical absorption spectrum in tandem with a conductance measurement, we could determine a diagnostic hole mobility value in minutes rather than hours. Maximizing this value became our optimization objective.

These types of diagnostic measurements enable fast comparisons between materials for the fast down-selection of materials. Promising candidate materials identified by the automated platform can then be fully characterized using conventional measurements.

4.6 Computer vision for flexible labs

Advances in computer vision are significant for flexible laboratory automation. Computer vision inspection provides both qualitative and quantitative data extraction with high precision and accuracy, even for systems that are nearly impossible for humans to resolve (e.g., very fast or slow processes, small features, invisible wavelengths) [38, 39]. Image analysis algorithms can now transform simple digital cameras into powerful analytical tools for laboratory use. It is striking how many properties relevant to the materials sciences (e.g., material defects and degradation, particle sizes and shapes, optical properties, film morphologies) can be evaluated with a simple digital camera and computer vision algorithms [26, 40–43]. A single vision module can thus supplant multiple application-specific instruments, allowing for a diverse array of both quantitative and qualitative property measurements to be performed with no new equipment. As part of an automated experiment, computer vision can both reduce the need for human supervision and observe phenomena incompatible with human perception. We are strong proponents of leveraging the speed and versatility of computer vision to more rapidly deploy automated experiments. [8, 26]

Consider the common experimental challenge in the materials science community of linking morphology and defects to the properties of a material. If, for example, we measure an increase in HTM hole mobility, is this change due to changes in dopant levels or film morphology? Answering this question is not trivial, but we were able to resolve these effects by adding imaging to our automated workflow. We used a camera with darkfield illumination to highlight morphological defects such as cracks and dewetting. Using these images, we were able to identify how the morphologies of organic thin films responded to variables such as dopants, additives, solvents, and thermal processing conditions [26].

A key benefit of including imaging in an automated workflow is that the large number of images obtained enable the training of machine learning algorithms. We used an image database generated by our HTM study to develop convolutional neural networks (CNNs) for quantifying morphological defects in images of thin films [26]. These CNNs enable our automated platform to perform rapid and autonomous image-based film morphology optimizations.

4.7 Robot/human interfacing

An automated system is most productive when it is easy to communicate with. For this reason, it is important to streamline both the programming of flexible automation platforms and the flow of data from the automation back to the experimentalist.

To simplify the programming of our automation platforms, we utilize the user-friendly open-source Python programming language, which has ready-built packages available for a wide variety of scientific tasks [44, 45]. As with our hardware, we focus on developing modular software that can be built and tested in isolation before being combined into larger automated workflows. This approach accelerates our development process by encouraging collaboration and code reuse. Programming new variations on our automated experiments in less than a day has become possible for our team as a result of this philosophy.

A notable benefit of automated platforms is that they can catalogue data about nearly every aspect of the experiments they perform [46]. It is difficult, however, for humans to quickly absorb the abundance of data from an automated platform. This mismatch can create a bottleneck which reduces the advantages of the automation.

We find tremendous value in using interactive data visualization "dashboards" that provide real-time analyses of automated data. These visualization tools provide a range of interactive plots and other information that enables the researcher to make on-the-fly assessments of how data fits a hypothesis. Dashboards also make it easy to query the details of every experiment. The researcher may, for example, identify reasons for the system to not perform a particular line of experiments. The ability to assess experiments on the fly can accelerate automated research workflows in a way that is difficult to achieve with manual experiments.

4.8 Flexible automation enables wider use of ML

Flexible automation presents an opportunity to enable the use of machine learning in research areas where this was previously not possible. While large empirical datasets are available in areas of materials science where high-throughput methods can be applied, such datasets are otherwise unavailable. Flexible automation can change this by making a broader set of traditionally manual experiments amenable to automation; this will increase the rate of data collection, and, in turn, the opportunities to employ machine learning.

Combining flexible automation and machine learning also provides opportunities to create new types of autonomous experiments or "self-driving laboratories" [5, 7–9, 47, 48]. These self-driving laboratories employ machine learning to discover new champion materials in fewer experiments[7, 49–50]. These algorithmic approaches to experiment planning can be effective even in cases where experimental throughput is limited, and are especially powerful when computational predictions are available [51]. Flexible automation is driving the emergence of a wide variety of self-driving laboratories for materials science. These systems will benefit from ongoing improvements in both flexible automation and in artificially intelligent tools for predicting materials properties, managing and analyzing experimental data, generating hypotheses, and planning experiments. Combining flexible automation with these machine learning tools to build effective self-driving laboratories is a promising new approach to accelerating the discovery of new materials.

References

[1]   Eagar TW. Bringing new materials to market. Technol Rev, 1995, 98, 42–49.
[2]   Häse F, Roch LM, Aspuru-Guzik A. Next-generation experimentation with self-driving laboratories. Trends Chem, 2019, 1, 282–291.

[3]   Stach E, DeCost B, Kusne AG, Hattrick-Simpers J, Brown KA, Reyes KG, Schrier J, Billinge S, Buonassisi T, Foster I, Gomes CP, Gregoire JM, Mehta A, Montoya J, Olivetti E, Park C, Rotenberg E, Saikin SK, Smullin S, Stanev V, Maruyama B. Autonomous experimentation systems for materials development: A community perspective. Matter, 2021, 4, 2702–2726.

[4]   Krishnadasan S, Brown RJC, deMello AJ, deMello JC. Intelligent routes to the controlled synthesis of nanoparticles. Lab Chip, 2007, 7, 1434–1441.

[5]   Nikolaev P, Hooper D, Webber F, Rao R, Decker K, Krein M, Poleski J, Barto R, Maruyama B. Autonomy in materials research: A case study in carbon nanotube growth. Npj Comput Mater, 2016, 2, 16031.

[6]   Bash D, Cai Y, Chellappan V, Wong SL, Yang X, Kumar P, Tan JD, Abutaha A, Cheng JJW, Lim Y-F, Tian SIP, Ren Z, Mekki-Berrada F, Wong WK, Xie J, Kumar J, Khan SA, Li Q, Buonassisi T, Hippalgaonkar K. Multi-fidelity high-throughput optimization of electrical conductivity in P3HT-CNT composites. Adv Funct Mater, 2021, 2102606.

[7]   Gongora AE, Xu B, Perry W, Okoye C, Riley P, Reyes KG, Morgan EF, Brown KA. A Bayesian experimental autonomous researcher for mechanical design. Sci Adv, 2020, 6, eaaz1708.

[8]   MacLeod BP, Parlane FGL, Morrissey TD, Häse F, Roch LM, Dettelbach KE, Moreira R, Yunker LPE, Rooney MB, Deeth JR, Lai V, Ng GJ, Situ H, Zhang RH, Elliott MS, Haley TH, Dvorak DJ, Aspuru-Guzik A, Hein JE, Berlinguette CP. Self-driving laboratory for accelerated discovery of thin-film materials. Sci Adv, 2020, 6, eaaz8867.

[9]   Li J, Li J, Liu R, Tu Y, Li Y, Cheng J, He T, Zhu X. Autonomous discovery of optically active chiral inorganic perovskite nanocrystals through an intelligent cloud lab. Nat Commun, 2020, 11, 2046.

[10] Langner S, Häse F, Perea JD, Stubhan T, Hauch J, Roch LM, Heumueller T, Aspuru-Guzik A, Brabec CJ. Beyond ternary OPV: high-throughput experimentation and self-driving laboratories optimize multicomponent systems. Adv Mater, 2020, 32, e1907801.

[11] Shimizu R, Kobayashi S, Watanabe Y, Ando Y, Hitosugi T. Autonomous materials synthesis by machine learning and robotics. APL Mater, 2020, 8, 111110.

[12] Ament S, Amsler M, Sutherland DR, Chang M-C, Guevarra D, Connolly AB, Gregoire JM, Thompson MO, Gomes CP, Van Dover RB. Autonomous synthesis of metastable materials. arXiv [cond-mat.mtrl-sci], 2021, at <http://arxiv.org/abs/2101.07385>.

[13] MacLeod BP, Parlane FGL, Dettelbach KE, Elliott MS, Rupnow CC, Morrissey TD, Haley TH, Proskurin O, Rooney MB, Taherimakhsousi N, Dvorak DJ, Chiu HN, Waizenegger CEB, Ocean K, Berlinguette CP. Advancing the Pareto front using a self-driving laboratory. arXiv [cond-mat.mtrl-sci], 2021, at <http://arxiv.org/abs/2106.08899>.

[14] Cao L, Russo D, Felton K, Salley D, Sharma A, Keenan G, Mauer W, Gao H, Cronin L, Lapkin AA. Optimization of formulations using robotic experiments driven by machine learning DoE. Cell Rep Phys Sci, 2021, 2, 100295.

[15] Maier WF, Stöwe K, Sieg S. Combinatorial and high-throughput materials science. Angew Chem Int Ed Engl, 2007, 46, 6016–6067.

[16] Potyrailo R, Rajan K, Stoewe K, Takeuchi I, Chisholm B, Lam H. Combinatorial and high-throughput screening of materials libraries: Review of state of the art. ACS Comb Sci, 2011, 13, 579–633.

[17] Green ML, Choi CL, Hattrick-Simpers JR, Joshi AM, Takeuchi I, Barron SC, Campo E, Chiang T, Empedocles S, Gregoire JM, Kusne AG, Martin J, Mehta A, Persson K, Trautt Z, Van Duren J, Zakutayev A. Fulfilling the promise of the materials genome initiative with high-throughput experimental methodologies. Appl Phys Rev, 2017, 4, 011105.

[18] Bloss R. Collaborative robots are rapidly providing major improvements in productivity, safety, programing ease, portability and cost while addressing many new applications. Ind Rob, 2016, 39, 88.

[19] Gasparetto A, Scalera L. From the Unimate to the Delta Robot: The Early Decades of Industrial Robotics. In: Explorations in the History and Heritage of Machines and Mechanisms, Springer International Publishing, 2019, 284–295.

[20] Kuo T-C, Malvadkar NA, Drumright R, Cesaretti R, Bishop MT. High-throughput industrial coatings research at the dow chemical company. ACS Comb Sci, 2016, 18, 507–526.

[21] Cohen AE, Ellis PJ, Miller MD, Deacon AM, Phizackerley RP. An automated system to mount cryo-cooled protein crystals on a synchrotron beam line, using compact sample cassettes and a small-scale robot. J Appl Crystallogr, 2002, 35, 720–726.

[22] Matheson E, Minto R, Zampieri EGG, Faccio M, Rosati G. Human–robot collaboration in manufacturing applications: A review. Robotics, 2019, 8, 100.

[23] Lochmüller CH, Lung KR, Cousins KR. Applications of optimization strategies in the design of intelligent laboratory robotic procedures. Anal Lett, 1985, 18, 439–448.

[24] Lochmüller CH, Lung KR. Applications of laboratory robotics in spectrophotometric sample preparation and experimental optimization. Anal Chim Acta, 1986, 183, 257–262.

[25] Coley CW, Thomas DA 3rd, Lummiss JAM, Jaworski JN, Breen CP, Schultz V, Hart T, Fishman JS, Rogers L, Gao H, Hicklin RW, Plehiers PP, Byington J, Piotti JS, Green WH, Hart AJ, Jamison TF, Jensen KF. A robotic platform for flow synthesis of organic compounds informed by AI planning. Science, 2019, 365.

[26] Taherimakhsousi N, MacLeod BP, Parlane FGL, Morrissey TD, Booker EP, Dettelbach KE, Berlinguette CP. Quantifying defects in thin films using machine vision. npj Comput Mater, 2020, 6, 111.

[27] Burger B, Maffettone PM, Gusev VV, Aitchison CM, Bai Y, Wang X, Li X, Alston BM, Li B, Clowes R, Rankin N, Harris B, Sprick RS, Cooper AI. A mobile robotic chemist. Nature, 2020, 583, 237–241.

[28] Wagner J, Berger CG, Du X, Stubhan T, Hauch JA, Brabec CJ. The evolution of materials acceleration platforms: Toward the laboratory of the future with AMANDA. J Mater Sci, 2021, 56, 16422–16446.

[29] Du X, Lüer L, Heumueller T, Wagner J, Berger C, Osterrieder T, Wortmann J, Langner S, Vongsaysy U, Bertrand M, Li N, Stubhan T, Hauch J, Brabec CJ. Elucidating the full potential of OPV materials utilizing a high-throughput robot-based platform and machine learning. Joule, 2021, 5, 495–506.

[30] Liang J, Xu S, Hu L, Zhao Y, Zhu X. Machine learning accelerated discovery of polymers through autonomous intelligent lab. Mater Chem Front, 2021, doi: 10.1039/D0QM01093F.

[31] Häse F, Roch LM, Kreisbeck C, Aspuru-Guzik A. Phoenics: a Bayesian optimizer for chemistry. ACS Cent Sci, 2018, 4, 1134–1145.

[32] Shiri P, Lai V, Zepel T, Griffin D, Reifman J, Clark S, Grunert S, Yunker LPE, Steiner S, Situ H, Yang F, Prieto PL, Hein JE. Automated solubility screening platform using computer vision. iScience, 2021, 24, 102176.

[33] Banks P. The microplate market past, present and future. Drug Discovery, 2009, 85.

[34] Hanak JJ. The 'multiple-sample concept' in materials research: synthesis, compositional analysis and testing of entire multicomponent systems. J Mater Sci, 1970, 5, 964–971.

[35] Xiang XD, Sun X, Briceño G, Lou Y, Wang KA, Chang H, Wallace-Freedman WG, Chen SW, Schultz PG. A combinatorial approach to materials discovery. Science, 1995, 268, 1738–1740.

[36] Koinuma H, Takeuchi I. Combinatorial solid-state chemistry of inorganic materials. Nat Mater, 2004, 3, 429–438.

[37] Blakesley JC, Castro FA, Kylberg W, Dibb GFA, Arantes C, Valaski R, Cremona M, Kim JS, Kim J-S. Towards reliable charge-mobility benchmark measurements for organic semiconductors. Org Electron, 2014, 15, 1263–1272.

[38] Shirmohammadi S, Ferrero A. Camera as the instrument: The rising trend of vision based measurement. IEEE Instrum Meas Mag, 2014, 17, 41–47.

[39] Capitán-Vallvey LF, López-Ruiz N, Martínez-Olmos A, Erenas MM, Palma AJ. Recent developments in computer vision-based analytical chemistry: A tutorial review. Anal Chim Acta, 2015, 899, 23–56.

[40] Zhang Y, Liu JJ, Zhang L, De Anda JC, Wang XZ. Particle shape characterisation and classification using automated microscopy and shape descriptors in batch manufacture of particulate solids. Particuology, 2016, 24, 61–68.

[41] Kobayashi Y, Morimoto T, Sato I, Mukaigawa Y, Tomono T, Ikeuchi K. Reconstructing shapes and appearances of thin film objects using RGB images. In Proceedings of the IEEE Conference on Computer Vision and Pattern Recognition, 2016, 3774–3782.

[42] Wieghold S, Liu Z, Raymond SJ, Meyer LT, Williams JR, Buonassisi T, Sachs EM. Detection of sub-500-μm cracks in multicrystalline silicon wafer using edge-illuminated dark-field imaging to enable thin solar cell manufacturing. Sol Energy Mater Sol Cells, 2019, 196, 70–77.

[43] Boldrighini P, Fauveau A, Thérias S, Gardette JL, Hidalgo M, Cros S. Optical calcium test for measurement of multiple permeation pathways in flexible organic optoelectronic encapsulation. Rev Sci Instrum, 2019, 90, 014710.

[44] Perkel JM. Programming: pick up python. Nature, 2015, 518, 125–126.

[45] Pérez F, Granger BE, Hunter JD. Python: an ecosystem for scientific computing. Comput Sci Eng, 2010, 13, 13–21.

[46] Pendleton IM, Cattabriga G, Li Z, Najeeb MA, Friedler SA, Norquist AJ, Chan EM, Schrier J. Experiment specification, capture and laboratory automation technology (ESCALATE): A software pipeline for automated chemical experimentation and data management. MRS Commun, 2019, 9, 846–859.

[47] Duros V, Grizou J, Xuan W, Hosni Z, Long D-L, Miras HN, Cronin L. Human versus robots in the discovery and crystallization of gigantic polyoxometalates. Angew Chem Int Ed Engl, 2017, 56, 10815–10820.

[48] Roch LM, Häse F, Kreisbeck C, Tamayo-Mendoza T, Yunker LPE, Hein JE, Aspuru-Guzik A. ChemOS: An orchestration software to democratize autonomous discovery. PLoS One, 2020, 15, e0229862.

[49] Cao B, Adutwum LA, Oliynyk AO, Luber EJ, Olsen BC, Mar A, Buriak JM. How to optimize materials and devices via design of experiments and machine learning: demonstration using organic photovoltaics. ACS Nano, 2018, 12, 7434–7444.

[50] Ren Z, Oviedo F, Thway M, Tian SIP, Wang Y, Xue H, Perea JD, Layurova M, Heumueller T, Birgersson E, Aberle AG, Brabec CJ, Stangl Rolf, Li Q, Sun S, Lin F, Peter IM, Buonassisi T, Embedding physics domain knowledge into a Bayesian network enables layer-by-layer process innovation for photovoltaics. npj Comput Mater, 2020, 6, 9.

[51] Gongora AE, Snapp KL, Whiting E, Riley P, Reyes KG, Morgan EF, Brown KA. Using simulation to accelerate autonomous experimentation: A case study using mechanics. iScience, 2021, 24, 102262.

John Dagdelen, Alex Dunn

5 Algorithms for materials discovery

5.1 Introduction

The problem of materials discovery is principally one of "finding a needle in a haystack"; among many possible configurations of atoms, which ones have technologically useful properties? To gain perspective on the scale of this problem, consider the space of quaternary stoichiometries represented by the formula $A_wB_xC_yD_z$ with $w,x,y,z \leq 8$. If we compute the number of combinations from the first 103 elements of the periodic table, there are more than 10^{12} possible stoichiometries. Even if we restrict this space based on well-known physical rules such as charge neutrality and electronegativity balance, our space still contains 32.3 billion possible stoichiometries [1]. This is an intractably massive space for exploration when we consider that the timeline for taking a single new material from theory to market is as long as 15–20 years [2]. Moreover, fundamental materials properties are determined not only by stoichiometry, but by atomic configuration (i.e. structure). For one stoichiometry, there are many potential atomic geometries, yet only a few energetically stable arrangements, and even fewer arrangements which are both stable and have desirable properties. Thus, considering only a miniscule fraction of all possible materials – single crystal quaternary compounds – we are met with an incomprehensibly large haystack of possibilities.

These massive-yet-sparse spaces of materials have been made less intimidating by advances in high throughput experimentation, data storage, and computational modeling. The "cheapest" and most plentiful datasets of materials properties are typically those generated with *ab initio* methods such as density functional theory (DFT), though large databases of experimental properties exist (Pearson Crystal DB/Pauling File [3, 4]). The largest databases of *ab initio* properties such as the open quantum materials database (OQMD) [5, 6], the Materials Project (MP) [7, 8], and the automatic-FLOW for materials discovery (AFLOW) [9] contain on the order of 10^5–10^6 unique materials. The most common properties for these materials stored in these repositories are the relaxed structures, total energies, and low-resolution band structures computed with density functional theory. These calculations are useful in determining predicted x-ray diffraction peaks, thermodynamic stability, and moderate fidelity estimates of electronic structure, but are often insufficient to determining many practical downstream properties or figures of merit of interest. Those material properties usually require more computationally intense methods, such as dielectric properties computed with density functional perturbation theory, mechanical properties such as elastic tensors calculated with multiple strained structures, and more accurate band structures computed

John Dagdelen, Alex Dunn, Univeristy of California Berkeley and Lawrence Berkeley National Laboratory, USA

https://doi.org/10.1515/9783110738087-005

with hybrid functionals. These properties are therefore far sparser in *ab initio* datasets than total energy calculations. For example, between all three of the repositories above, there are less than 30,000 DFT-computed elastic tensors.

Applying machine learning (ML) models to materials data presents an opportunity to speed up materials screening and discovery by orders of magnitude. Whereas computing the elastic moduli of 1,000 crystals with DFT may require weeks of computation across thousands of cores on supercomputers, a trained ML model can provide moderate fidelity estimates for the same quantity in an hour on a modern laptop [10]. ML property prediction algorithms can also address questions not easily answered by *ab initio* methods or molecular dynamics (MD) simulations, such as determining properties of disordered structures (e.g., glasses).

Aside from property prediction, ML models may supplement canonical knowledge by providing insight into scientific phenomena and practical procedures, helping to answer questions like:

– Are there new candidates for thermoelectric generators among previously synthesized materials which have not yet been investigated as thermoelectric?
– What are some interesting chemical systems to explore for novel bulk metallic glasses?
– What is the most reliable route for synthesis of high-quality Ag-doped $SrTiO_3$?

In this chapter, we outline the general procedures for using machine learning in materials discovery and provide some examples of using machine learning to accelerate the identification and development of technologically relevant materials candidates. Next, we outline the basic mathematical tenets of machine learning algorithms most commonly applied to materials discovery problems. Finally, we examine the contributions of natural language processing (NLP) to materials discovery and outline opportunities for further exploration.

5.2 The funnel for materials discovery

The main computational materials screening workflow today is the "funnel" strategy (Fig. 5.1.) Many initial candidates are considered and less-promising candidates are screened out at successive levels (representing various levels of predictive fidelity) of the funnel. Screening funnels have conventionally been based on *ab initio*density functional theory or molecular dynamics simulations for property predictions, but machine learning is increasingly being used as a valuable screening stage in these workflows. The stages of the screening funnel including machine learning are loosely as follows:

1. Identify a (large) screening space of potential candidates.
2. Train a machine learning model on available data to predict missing properties of interest.

3. Screen based on ML predictions of the desired properties of interest.
4. Run high-throughput DFT or MD computations on the most promising candidates from the previous step. For further investigation or validation, run higher order methods (e.g., all electron methods, more expensive DFT procedures) to confirm expected properties.
5. Experimentally synthesize the most promising candidates.

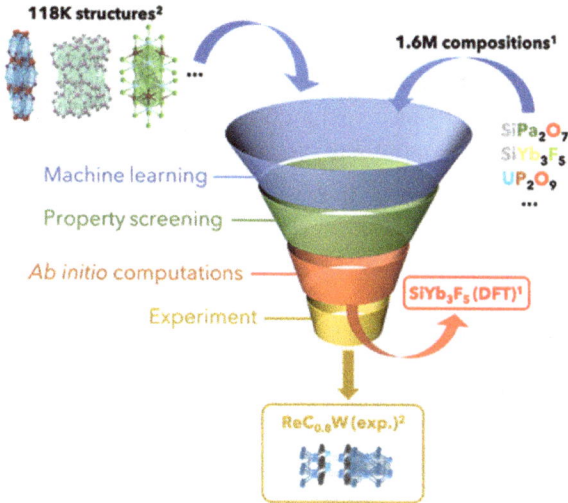

Fig. 5.1: The funnel approach to materials discovery, with examples for (1) novel stable compositions [11] and (2) superhard materials [12].

All screening funnels start out with some database containing many possible candidate structures, such as the Materials Project. The goal of the funnel is to sequentially apply filters that screen out structures that either have unsuitable properties for the desired application or would otherwise be difficult to employ as the active material in a device. The first screening stages in the funnel utilize whatever precomputed properties are available to narrow down the number of candidates as much as possible. These properties can include quantities such as band gap, composition, lattice constant, and formation energy, which are all readily available in the major materials databases. The most important of these criteria is plausible synthesizability and stability, which is correlated with low formation energy relative to neighboring compounds in composition space. Cost criteria can also be estimated from composition to screen out compounds containing rare and expensive elements.

However, what if we wish to screen materials based on properties that have only been calculated for a small portion of our candidate pool? In these cases, it is often overly expensive or impossible to calculate or measure these properties for every remaining candidate. Instead, we can train a machine learning model on the limited data available and estimate the desired property for the entire candidate pool. By selecting materials on the basis of these predictions, we can limit more computationally expensive calculations or time-consuming experiments to a smaller number of compounds that are more likely to have the desired properties and be plausible for real applications.

The ultimate goal of any materials screening funnel is to bring cheaper, functional, environmentally conscious candidates all the way to mass production in industry. However, intermediate outputs of materials discovery pipelines can also be massively impactful; laboratory synthesis of novel performant compounds is a significant landmark toward deployment of those compounds in industry. Further, even the computational confirmation of novel compounds with *ab initio* methods represents a stepping-stone for further investigation into yet-undiscovered scientific phenomena, synthesis methods, and optimization.

5.3 Examples of using machine learning for materials discovery

The application of ML algorithms to materials discovery has grown enormously over the past decade. As shown in Fig. 5.2, the number of scholarly publications (as estimated by the Web of Science) matching both "machine learning" and "materials science" has grown by more than two orders of magnitude over the past 15 years. A few select (and by no means exhaustive) set of applications machine learning has been applied to are photocatalysts [13], thermoelectric generators [14], bulk metallic glasses [15], photovoltaics [16, 17], magnetically and topologically ordered materials [18], battery cathodes [19], ferroelectrics [20], high-entropy alloys [21], and superconductors [113, 114].

While a comprehensive review of all areas of materials discovery which have been accelerated by ML is outside the scope of this chapter, we will very briefly introduce two illustrative examples. Interested readers can also find a number of extensive review papers that have been written on the topic with additional examples [23–27].

Scientific publications per year mentioning both 'machine learning' and 'materials science'

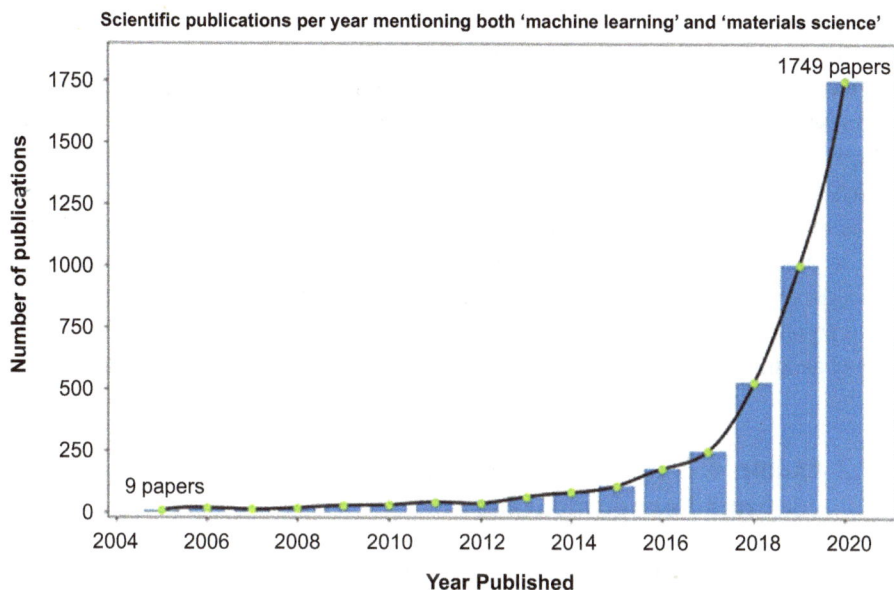

Fig. 5.2: Estimate of number of machine learning + materials discovery papers published, by year (2005–2020) according to the Web of Science.

5.3.1 Example 1: support vector machines help experimentally realize novel superhard materials

Tehrani et al. [12] developed a machine learning approach for predicting likely superhard (Vickers microhardness measurements ≥40 GPa) materials. Conventionally predicting material hardness is difficult, as it is a property dependent on mechanical phenomena across length scales including crystal structure as well as microstructure. Tehrani et al. utilized the elastic moduli (bulk modulus K and shear modulus G) as proxies for hardness, defining a valuable metric which a machine learning model could in theory predict. Using the 3,428 elastic moduli (calculated with the Voight–Reuss–Hill approximation [28], K_{VRH} and G_{VRH}) already in the Materials Project database cross-referenced with ambient pressure/temperature phase data from the experimental Pearson Crystal Database [3], their investigation fits a support vector machine (SVM) machine learning model to predict the elastic moduli of 118,287 compounds. Their SVM, which employed a radial basis function as a kernel to model nonlinear dependencies, was coupled with a genetic algorithm for feature selection. The final version of their model was able to predict bulk moduli in their validation sets within ±20% of the DFT-computed versions. Using their K and G predictions, Tehrani et al. identified and synthesized $ReWC_{0.8}$ and $Mo_{0.9}W_{1.1}BC$ via arc melting. Both of these compounds, never before identified as superhard candidates, were measured via micro indentation and confirmed with Vickers hardness measurements above 40 GPa under low loads.

Key Takeaways. Tehrani et al.'s relatively simple SVM approach demonstrates the massive opportunities that machine learning presents in the materials domain. Whereas performing DFT calculations to screen the entire 118,000+ candidate search space is infeasible – requiring many months or years running across thousands of high-performance compute cores – their investigation leveraged existing data to predict likely candidates using relatively simple and inexpensive statistical methods. Moreover, although their SVM model was certainly imperfect (having root mean squared errors of 18.4 GPa for G_{VRH} and 14.3 GPa for K_{VRH}), their model was indeed accurate enough to identify viable candidates for experimental validation.

5.3.2 Example 2: novel Heusler structures suggested by ML are confirmed experimentally

Oliynyk et al. [29] used a machine learning model to identify and synthesize entirely novel Heusler compounds, a structure class which has formulas AB_2C and exhibits interesting thermoelectric, spintronic, and magnetocaloric properties. Beginning from a space of over 400,000 potential candidates, their investigation represented compounds as sets of 22 features – including chemically relevant information, such as electronegativity values, valence electron counts, and atomic radii differences of atoms on the A and B sites – for input to a random forest regression algorithm. Their algorithm was trained on existing data from confirmed AB_2C compounds. Finally, using predictions of the probability that a given AB_2C composition crystallizes in the Heusler structure, they synthesized two series of then-unknown gallides, MRu_2Ga and RuM_2Ga (M = Ti – Ni). All the synthesized compounds except for the M = Ni variant were confirmed as Heusler structures via powder X-ray diffraction.

Key takeaways. Determining stable structures directly from stoichiometries is an often arduous and uncertain process given the many possible arrangements of atoms and complex energy surfaces involved in finding stable compounds. Focusing on predicting the probability that a specific family of stoichiometries matches a particular structure prototype (Heuslers) considerably narrows the problem to a feasible machine learning experiment. Not only were several *families* of previously unknown structures experimentally validated, they were selected from a search space larger than any existing experimental setup (and all but the largest *ab initio* computing centers) would be capable of exploring.

5.4 A note on features/descriptors/representations

Any machine learning model is only as good as the inputs it receives. Thus, creating mathematical representations (also known as features or descriptors) that encode the relevant physical and chemical information for materials is vital. In most studies, these features are derived from chemical composition (e.g., weighted statistics on properties of elements such as the popular "Magpie" set introduced by Ward et al [30]), crystal symmetry, local order/nearest neighbor environment, electron configuration, or microstructural properties. The standard features are often also supplemented by hand-crafted descriptors selected specifically for the task at hand; for example, in predicting the stability of perovskite structures the Goldschmidt tolerance factor [31] is a pertinent specialized descriptor.

In recent years, the conventional process of manual feature engineering has been challenged by "featureless" deep neural networks [32–36] which can internally determine optimal descriptors. These networks learn representations directly from chemical composition or crystal structure and can exceed the performance of manually engineered feature sets, especially on large datasets and specific properties (e.g., formation enthalpy) [37].

We will not discuss manual feature engineering as it varies with almost every study and there are dozens of methods that we could discuss, many of which are specific to certain subdomains or classes of materials. Some review papers have covered this topic in good detail and are very useful references for those getting started in the field [27]. In the later sections, we will briefly discuss how deep neural networks can create internal sets of features – or embeddings – which are useful for a variety of learning tasks.

5.5 What we haven't covered in this chapter

A practitioner of machine learning must be cognizant of, perhaps obsessed with, the validation of their models. In essence, we should estimate how well a model will perform on real data outside of the training set. Validating and testing models is critical for understanding model performance, but while we will not cover validation procedures in this chapter, there are many excellent resources on the subject, such as Hastie et al.'s *Elements of Statistical Learning* [38].

In the remainder of this chapter, we present a survey of some of the core machine learning algorithms used for materials discovery today, including some newer algorithms that are likely to become increasingly relevant for materials discovery in the near future. However, the reader may note that a fair number of relevant algorithms are not discussed or that some are described in insufficient detail. Unfortunately, a full discussion of every major machine learning algorithm would warrant entire books on the subject (indeed there are many ML books to choose from [38, 39].) Our hope with this chapter is that materials scientists new to the field may find it a useful introduction that can help them on their materials discovery journey.

5.6 Model-based machine learning

The goal of model-based machine learning algorithms is to construct a mathematical function, or *model*, that captures patterns from data in such a way that allows us to make useful predictions/inferences about new data points or understand the existing data set in a new light. At a high level, all machine learning algorithms follow the same general structure.

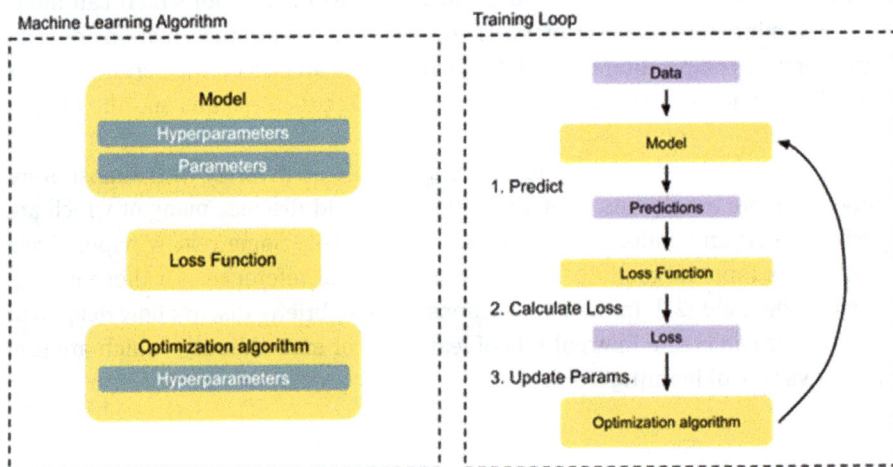

Fig. 5.3: Overview of the components of a machine learning algorithm and the training loop.

ML models take in an *input*, which is a standardized representation of a single instance sampled from set of data, and yield an *output* prediction, which can take several forms but is usually a scalar or vector quantity. To construct these models, one first defines a *loss function*, which expresses a scalar representation of a given model's performance on a desired task as a function of the model, input, and label associated with that input. To be trainable, machine learning models must be *parameterized*, which means the operations they perform depend on a set of tunable *parameters* (also referred to as *weights*.) During training, the model's parameters are iteratively adjusted to optimize the model's performance (i.e. the loss function score) by repeatedly evaluating the model on a set of *training data* and adjusting the parameters so as to improve its performance via some kind of *optimization algorithm* (Fig. 5.3, right). ML algorithms can also include a set of *hyperparameters* which define aspects of the model architecture and learning process. While parameters are learned by the algorithm during training, hyperparameters are selected before training (either by the practitioner or by an automated meta-procedure). Common hyperparameters control the number of parameters in the machine learning model, the number of examples used during each loop of the parameter update procedure (batch size), and the step size for the parameter updates (learning rate). The training process is often repeated multiple times with various values for these hyperparameters to determine the optimal set for the problem at hand.

Using this general structure, different machine learning algorithms can easily be compared to each other according to these component parts, assembled like Lego® bricks (Fig. 5.3, left). New machine learning algorithms are constructed by modifying or replacing one or more of these components.

5.7 Tree-based models, random forests, and XGBoost

Tree-based models are historically some of the most successful ML architectures for materials discovery tasks because they perform well on small to medium datasets, tend to over-fit less, and can be more interpretable than other ML techniques. They have been used to discover new ternary solids [11, 29], bulk metallic glasses [30, 40], and Ni-rich Li-ion battery cathodes [41].

All tree models, including gradient boosted trees and random forests, are fundamentally made up of one or more *decision trees*, similar to "flowcharts" to aid in decision making. The key difference in machine learning decision tress is that the splits are not determined *a priori*, but rather in response to a dataset.

5.7.1 Decision trees

At each branch node in a decision tree, a binary split is made based on a single feature. The branches of trees eventually terminate at *leaf nodes*, which output the label predicted for the input data. The main hyperparameter of a decision tree model is the *maximum depth*, which sets the maximum number of branches that are made. Leaf nodes are be defined by at least one data point from the training set, so the dataset size also affects the maximum theoretical depth for the tree. Increasing the maximum depth tends to decrease the bias of a model (under-fitting) and can improve performance, however increasing depth can also increase the variance of the model and result in over-fitting. If a large maximum depth is used, in the extreme case, a tree can be constructed where every leaf node contains a single data point, which is usually going to be over-fit and generalize poorly.

Another reason decision trees are popular is that they can be more directly interpretable because they make explicit binary comparisons at each node, which can be interrogated analytically.

5.7.1.1 Building trees for classification

In classification tasks, splits are chosen so as to maximize the *information gain* of the split, which is defined as the difference in entropy, $H(S)$, between the original set of data points and the average entropy of the two new sets of data after the split:

$$H(S) - \frac{|S_1|}{|S|}H(S_1) - \frac{|S_2|}{|S|}H(S_2)$$

The information entropy, $H(S)$, is analogous to the notion of configurational entropy in statistical mechanics/thermodynamics and is defined as

$$H(S) = -\sum_{c=1}^{C} p(c) \log p(c)$$

where c is the class label in the set $C = \{c_1, c_2, \ldots, c_n\}$ and $p(c)$ is the proportion of the data belonging to class c. The practical effect of this loss function is to reward splits that lead to data from the same class collecting on the same side of the split. Branching is stopped when all the samples at a leaf belong to the same class, when a minimum number of samples occurs at a leaf, or when the maximum depth is reached.

5.7.1.2 Building trees for regression

In regression tasks, splits are typically chosen so as to maximize the *standard deviation reduction* of the split. For each potential branch condition, we compare the standard deviation of the resulting split to the standard deviation of the parent node

$$\sigma_{\text{reduction}} = \sigma_{\text{parent node}} - \sigma_{\text{potential split}}$$

where the standard deviation of any sample is defined as:

$$\sigma = \sqrt{\frac{\sum_{i}^{n} (x_i - \bar{X})^2}{n}}$$

where n is the number of samples, $\bar{X}$ is the sample mean, and x_i is the ith sample value. The σ of the parent node is the standard deviation of the target values (i.e., the value you want to predict) in the parent node. The σ of the potential split is the sum of the standard deviations of potential split outcomes weighted by their probabilities, which is analogous to the definition of expectation values in statistical mechanics:

$$\sigma_{\text{potental split}} = \sum_{o \in O} \frac{n_o}{n} \cdot \sigma_o$$

where O is the set of all outcomes (there are only two if the feature is continuous, though many if the feature is categorical), o is a particular outcome $o \in O$, n is the total number of samples considered among all outcomes, n_o is the number of samples in outcome o, and σ_o is the standard deviation of the target values in outcome o.

Splits are chosen based on the maximum standard deviation reduction (similarly referred to as "sum of squared errors", SSE) among all the potential splits considered. Branching ends once a node has a minimum number of samples, a minimum standard deviation relative to the number of samples in the split is reached, or the maximum depth is reached. In general, regression trees map new data points onto a set of discrete possible values, not a continuous output. Predictions for new data points are derived by following the decision tree until a leaf node is reached and then using the mean of the data points in the leaf nodes is returned as the prediction. If there is only one data point in the leaf note, the value of that data point is used as the prediction.

5.7.2 Bagging

One way to lower the variance of predictions is to train multiple models, known as an *ensemble*, and average their predictions. However, constructing decision trees is deterministic with respect to the training data, so the only way to generate and ensemble of models with any diversity is to generate them with different training sets. Bootstrap aggregation, or "bagging," is the process of combining predictions from an ensemble of models fit on subsets of the training data in order to produce a more robust model and reduce over-fitting [42]. Using bagging with decision trees gets around the issue of identical trees and allows one to train an ensemble of decision trees. However, this is often not enough to achieve models that generalize well because it's common for a few of the features to dominate the initial splits in decision trees. When this is the case, the resulting trees are not very diverse and ensembling them does not reduce variance much. Luckily, this problem is solved by the next tree-based model in our discussion: random forests.

5.7.3 Random forests

Random forests solve the diversity issue by taking the bagging approach further and bagging over both training points and input features [43]. These models are made up of ensembles of decision trees where each tree only has access to a random subset of features and only considers a random subset of the training data. This results in models that are harder to over-fit and generally results in trained models that generalize better. As a result, in materials discovery applications where datasets are often small (fewer than 10,000 points) and neural networks are prone to over-fitting, random forests are often a great choice. Random forests have been used widely for the discovery of new materials.

An early success story of modern materials ML by Meredig et al. showed that random forests could be used to discover new ternary compounds [11]. The authors trained a random forest model on DFT formation energies and a set of 129 descriptors (not derived from DFT) for each compound in a set of 4,000 ternaries and used it to screen 1.6 million candidate ternary compositions for stability. They identified 4,500 compositions likely to have unknown stable crystalline phases and then performed subsequent DFT calculations to confirm this to be the case for a number of the compositions. While newer models for formation energy prediction from composition have been developed in recent years, random forests can still often provide competitive performance on small datasets and are usually much simpler to train.

Because of this, random forests are often used as baseline models for comparing new models against, and packages for training random forests are well-developed and are extensively available in libraries including scikit-learn [44] and automatminer [37].

5.7.4 Gradient boosting

Another approach to reducing the variance of decision trees is to use a technique called *gradient boosting*. The key idea of gradient boosting is that the variance of a decision tree can be reduced by training another decision tree to estimate a correction term that make its predictions more accurate. By chaining these correction trees together, we can build up a model with improved generalization characteristics. XGBoost [45] is an efficient implementation of this algorithm that is very well optimized. An example of XGBoost being used for materials discovery comes from Schleder et al. [46] who trained an XGBoost model to predict the electronic topology of 2D materials with an accuracy of over 90% and used it to discover 17 new topological insulators.

5.8 Optimization

The goal of any optimization algorithm is to find the minima or maxima of a particular function. In the introduction, we alluded to a general structure of training ML models by feeding a parameterized loss function to an optimization algorithm. Before we get into the rest of the machine learning algorithms presented in this chapter, we will take a quick detour to learn about the most common optimization technique used in these algorithms, gradient descent, and a machine learning algorithm for directly optimizing materials properties, Bayesian optimization.

5.8.1 Stochastic gradient descent

Gradient descent is by far the most common type of optimization algorithm used to train machine learning models [47]. The idea behind gradient descent is to iteratively calculate the partial derivatives (or gradient) of the loss function and update the parameters of your model to take a "step" in the direction of the calculated gradients. This step size is a hyperparameter known as the "learning rate." Larger learning rates result in faster training, but learning rates that are too high will oscillate around solutions and fail to converge.

When the loss function is convex (has a single minimum), differentiable, and continuous, gradient descent is proven to converge to the optimal solution as the step size approaches 0 [48]. However, in materials ML, non-convex loss surfaces are ubiquitous. In these cases, gradient descent can get stuck in local minima and converge to sub-optimal solutions. Moreover, using gradient descent can be slow because the partial derivatives of the loss function have to be calculated for every data point in the dataset for each iteration, which is very computationally intense.

To overcome this issue, we can use methods like stochastic gradient descent (SGD) or mini-batch stochastic gradient descent instead [47]. In SGD, during each iteration the partial derivatives of the loss function are only calculated for a randomly chosen subset of the data at each iteration. This results in a noisy path that can help push the optimizer out of local minima. While a larger number of iterations is required to converge to a solution when using SGD, the individual iterations complete much faster than calculating derivatives for every sample in the dataset and the overall training wall time in greatly reduced. While in normal SGD, only a single randomly selected sample is used, in mini-batch SGD, multiple are used and their gradients are averaged. This affords the user a second hyperparameter, the mini-batch size, that can be used to fine-tune the training process.

In addition to SGD, there are numerous other optimization algorithms (mostly based on SGD) available with various mechanisms that help the optimizer escape local minima. The Adam optimizer [49] is one of such optimizers and is currently a popular choice for training neural networks.

5.8.2 Active design/Bayesian optimization

While optimizers like stochastic gradient descent are used for updating the parameters of machine learning models, there are also optimization strategies which can be directly used for materials discovery. Let's say we want to find materials with a high value of some "objective function" such as high turnover frequency of a heterogenous catalyst, and we have a little bit of data to train a machine learning model on but not enough to train a really accurate model. However, we do have the ability to collect more data, albeit using an expensive method.

This is known as a "black box" optimization problem. It's called "black box" because we don't know the inner workings of the objective function that we want to maximize. We can enter inputs into the black box and get values out, but we can't calculate gradients for its output with respect to our inputs. Moreover, we have an *expensive* black box so it's not feasible to just brute force the problem by trying every possible candidate.

Fortunately, there is a wealth of research on these kinds of problems and we can employ an "in the loop" strategy that combines machine learning and conventional experiments/calculations to accelerate our materials discovery task:

1. Train a model on available data from accurate, but expensive experiments/calculations.
2. Use the model to determine the next best candidate to explore.
3. Evaluate the candidate with the more accurate method.
4. Update the dataset with the results from the experiment/calculations.
5. Repeat

This general iterative framework is most commonly known in the materials discovery domains as *active learning*, *active design*, or *surrogate modeling*. The most common black-box optimization used for these frameworks is Bayesian optimization (BO). BO is an extraordinarily flexible tool for materials discovery. It has been applied to a very wide range of problems including exploring novel spaces of ternary nitrides and carbides [50], shape memory alloys [51], defect distributions [52], atomic clusters [53], and NiTi alloys [54]. There are also multiple software packages available for running various BO optimization loops specifically for materials discovery use cases [55–57].

Despite the wide applicability of BO, it is a relatively straightforward framework composed of only a few core parts. BO works by continually updating a probability distribution (the *prior distribution*, previous knowledge) with current knowledge to create a *posterior distribution*. Bayesian optimizers select new points using an *acquisition function*, which is a mathematical formalism that aims to balance exploration vs. exploitation based on the prior distribution. The most common acquisition function is called *expected improvement*, written as EI or $E[I(x)]$. EI is merely the expectation value of $I(\mathbf{x}) = \max(f_{min} - Y, 0)$, which represents the difference between the current known optimal value f_{min} and the predicted values of unknown points modeled as a random variable Y. Note that every prediction for an unknown value is a normal distribution around a mean $\hat{y}$ and standard error $s = \sigma/\sqrt{n}$. Therefore, to obtain the expected improvement EI, we can use the closed form equation:

$$E[I(x)] = (f_{\min} - \hat{y})\Phi\left(\frac{f_{\min} - \hat{y}}{s}\right) + s\phi\left(\frac{f_{\min} - \hat{y}}{s}\right)$$

where Φ is the Gaussian cumulative distribution function and ϕ is the Gaussian probability density function [58].

To predict the "next best point" in the exploration space, all we need is a model which returns predictions with distributions/uncertainty about a number of points that we are interested in exploring. The best point we can choose to explore next is the one with the maximum EI.

This leaves the question: *How do we predict unknown points with probability distributions to represent their uncertainty?* Bayesian optimization is most commonly used with a Gaussian process model [59] (also known as Kriging regression) since the model automatically returns predictions and probability distributions. However, by utilizing bagging (Section 5.7.2) and repeatedly training multiple models on random mixes of training data, we can use most machine learning models to create an ensemble of predictions to obtain a standard deviation σ and mean prediction $\hat{y}$ as surrogates to compute EI. With enough computation, you can use almost any of the models introduced in this chapter in a Bayesian optimization framework.

5.9 Linear models

We'll resume our overview of ML algorithms for materials discovery with the illustrative example of least squares linear regression (LR), which is an algorithm that many researchers are at least somewhat familiar with. Linear regression fits an N-dimensional hyperplane (a line in N-dimensional space) to a dataset by minimizing its average square distance to points in the dataset. In linear regression, the model is defined as a linear function that takes in a vector as an input, multiplies each entry in the vector by a corresponding weight, sums them together, and adds a bias term. In mathematical notation, this is expressed as

$$f(x_i) := \sum_{j=1}^{p} \beta_j x_{ij} + \beta_0 \tag{5.1}$$

where x_{ij} is the jth the entry in the input vector x_i, β_j is the weight corresponding to the jth entry, and β_0 is a bias term. The loss function of linear regression is the residual sum of squares (RSS), which is defined as:

$$\text{RSS} = \sum_{i=1}^{n} (y_i - f(x_i))^2 = \sum_{i=1}^{n} \left(y_i - \beta_0 - \sum_{j=1}^{p} \beta_j x_{ij}\right)^2 \tag{5.2}$$

LR is so common that convenient standalone functions for performing a least-squares linear regression are available in a number of python libraries including scikit-learn and scipy.

5.9.1 Regularization

The goal of regularization is to lower the variance of a model and thereby improve how it generalizes to unseen data. Regularization techniques often leverage insights about how models over-fit, such as by placing too much emphasis on certain features or using too many extraneous features. A common approach to regularize models is adding terms to the loss function that constrain the learned model in a way that reduces this behavior. For example, we can regularize the residual sum of squares loss used by linear regression by adding a regularization term that is a function of the model's weights:

$$\text{Regularized RSS} = \sum_{i=1}^{n} \left(y_i - \beta_0 - \sum_{j=1}^{p} \beta_j x_{ij} \right)^2 + R(\boldsymbol{\beta})$$

where $R(\boldsymbol{\beta})$ is this regularization term, which depends on the weights $\boldsymbol{\beta}$.

5.9.1.1 Ridge regression

Ridge regression adds a term to the loss function called the L_2-norm. This encourages models to learn smaller weights and not over-emphasize certain features

$$\text{Ridge Loss} = \text{RSS} + \lambda \sum_{j=1}^{p} \beta_j^2$$

5.9.1.2 LASSO

The least absolute shrinkage and selection operator, or LASSO, adds a slightly different term to the loss function that penalizes absolute values of the weights. This is known as the L_1-norm. In addition to favoring smaller weights, the L_1-norm encourages models use a few nonzero weights rather than many small weights.

$$\text{LASSO Loss} = \text{RSS} + \lambda \sum_{j=1}^{p} |\beta_j|$$

A notable example of using LASSO regression for materials discovery comes from Bucior et al., who used the method to predict hydrogen uptake in metal-organic frameworks to an accuracy within 3 gL^{-1} and the experimental verification of a new hydrogen storage material with a hydrogen deliverable capacity of 47 gL^{-1} [60].

5.9.2 Logistic regression

What if our task is to predict a binary yes/no labels for our data rather than scalar values? In this case, we can make a minor modification to our linear regression method by wrapping equation 5.1 with the *logistic* function

$$s(f(x)) = \frac{1}{1 - e^{-f(x)}}$$

where $s(f(\mathbf{x}))$ is the probability of the positive class label.

Logistic regression is a simple, yet powerful, classification technique for materials discovery. For example, Sendek et al. used logistic regression to predict whether a material would exhibit fast Li-ion conduction at room temperature from atomistic features of the material's unit cell [61]. They screened more than 12,000 Li-containing compounds in the Materials Project for thermodynamic phase stability, low electronic conduction, high electrochemical stability, and stability against reduction, and then used their logistic regression model to rank the candidates for Li conductivity. This combination of a conventional materials screening funnel and a machine learning model yielded four new solid Li-ion conductors, verified via DFT [62].

5.9.3 Kernel methods

Using a different basis for our data allows us to represent data in a higher dimensional vector space, which allows us to fit functional forms that are more complex. These functions are still linear in the "lifted" vector space, but can be non-linear in our original feature space. For example, we could use a new basis that includes every possible power of a feature (e.g. $w_1 x_i + w_1 x^2_i + w_3 x^3_i + w_4 x^4_i + \dots$). This new basis is known as a "kernel", in this case the polynomial kernel. The "kernel trick" is a technique that allows machine learning models to learn in this lifted, nonlinear basis without explicitly calculating the infinite number of non-linear features.

When combined with ridge regression, this yields a very useful machine learning technique known as *kernel ridge regression*. The most common kernels for kernel ridge regression are the polynomial kernel and the Gaussian kernel (also known as the radial basis function kernel or RBF kernel). Gaussian kernel ridge regression was used in combination with genetic algorithms by Mannodi-Kanakkithodi et al. to discover new polymer dielectrics that were then verified with DFT calculations [63].

Another common linear algorithm used with the kernel trick is the *support vector machine* (SVM), which can be used for both regression and classification. SVM creates a decision boundary known as a hyperplane (a line in N-dimensional space) which separates data into different classes. SVM utilizes a loss function which maximizes the *margin* between sets of points belonging to disparate classes and the hyperplane. It's convenient to think of the margin as a protected boundary between classes which the algorithm is trying to maximize. While SVM without a kernel is a linear method, SVM is typically used alongside radial basis functions or polynomial kernels.

5.10 Introduction to neural networks

Neural networks are quickly becoming one of the most active areas of research for materials discovery. Artificial neural networks (commonly referred to as simply "neural networks" in the literature) comprise a class of machine learning architectures inspired by structures in the brain.[2] Neural networks are very popular in modern materials machine learning because they can be extremely expressive and very often produce state of the art performance on a wide variety of tasks with less manual feature engineering.

5.10.1 Neurons

The fundamental building block of a neural network is a *neuron*, which takes in a vector of inputs and produces a scalar output. Multiple artificial neurons can be gathered together into *layers* that take in a single vector of inputs and yield a vector of outputs. Each ith neuron in a layer is parameterized by a set of weights $\mathbf{w}_i$ and a bias term b_i, which gets applied to the input vector $\mathbf{x}$ as

$$y_i = g(\mathbf{w}_i \cdot \mathbf{x} + b_i) = g\left(\sum_j w_{ij}x_j + b_i\right)$$

where x is the vector of inputs, y_i is the output of the ith neuron in the layer, and g is some function that transforms the output.[3] The reader might note this looks very similar to the equation for linear regression 1 other than the g function wrapping it. So, what is this g?

---

2 However, it should be noted that artificial neural networks and biological neural networks actually function very differently. This discussion is beyond the scope of this chapter, but there are good discussions readily available online.

3 In machine learning resources and codes this expression is often modified by moving the bias term inside of **w** and appending a 1 to the beginning of the feature vector **x**. in this case, the function becomes $y_i = g(\sum_j w_{ij}x_j)$.

5.10.2 Activation functions

The activation function, g, of an artificial neuron is the "secret sauce" that allows artificial neural networks to produce complex, nonlinear mappings between data and labels. Activation can introduce non-linearity by transforming the output of neurons by a non-linear function. Some of the most common activation functions used in machine learning are described in Table 1 (Table 1: Common activation functions used in neural networks):

Name	Function	Non-linear?
Identity	$g(x) = x$	No
Logistic	$g(x) = \dfrac{1}{1 + e^{-x}}$	Yes
ReLU	$g(x) = \begin{cases} 0, & \text{if } x < 0 \\ x, & \text{if } x \geq 0 \end{cases}$	Yes
Hyperbolic tangent	$g(x) = \tanh(x) = \dfrac{2}{1 + e^{-2x}} - 1$	Yes

5.10.3 The structure of neural networks

Artificial neural networks are constructed by assembling "layers" of neurons which all take in the same input and produce individual outputs. The number of neurons in a layer is called the layer's "width." Layers can be sequentially stacked together, each layer taking the outputs of the previous layer as its own input features. The number of layers in a neural network is known as the network's "depth" and layers in between the input and output layers are known as "hidden layers."

"Deep learning" is the sub-field of machine learning concerned with training neural networks with many hidden layers [64]. Until recently, it was impractical to train deep learning models due to both the computational cost and other issues that made optimizing deep neural networks practically difficult. However, today computational cost is no longer a barrier to training deep neural networks and techniques such as stochastic gradient descent, gradient clipping, skip-connections, and residual networks have made training deep networks with many layers feasible.

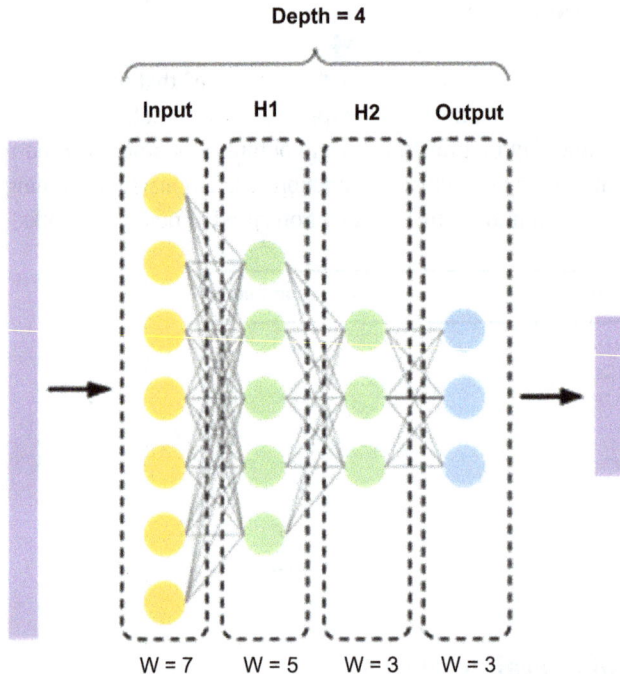

Fig. 5.4: Schematic of a four-layer deep neural network with two hidden layers, H1 and H2, of width 5 and 3, respectively. This neural network outputs a vector of size 3.

5.10.4 Regularizing neural networks

Like in the linear regression example we presented earlier, one can also regularize neural networks by adding extra terms to the loss function, such as a L1 or L2 norm. However, there are also other ways that neural networks are commonly regularized. For example, we can randomly select some neurons in the neural network at each training step and drop their inputs and outputs from the prediction [65]. This technique is known as *dropout* regularization, and it has been used extensively in deep learning.

The intuition behind dropout is that it breaks up situations where network layers rely too heavily on mistakes from prior layers. This prevents layers becoming detrimentally co-adapted and also has the side effect of causing the activations of the hidden units to become sparser, even without L1 or L2 terms in the loss function. The overall effect of this change can make neural networks more robust to unseen data.

Another approach to reduce over-fitting in neural networks is to train the neural network on multiple copies of each data point with minor modifications that don't affect the label that should be predicted, such as rotating or translating images. Data augmentation is mostly used in image recognition and computer vision, but it can be applied to other problems as well. Intuitively, data augmentation can teach a model about invariances in the data domain that don't affect a data point's label. Examples include cropping or shearing images or different orientations of the same crystal structure in 3D space.

5.10.5 Invariant and equivariant neural networks

For most materials property prediction tasks, the prediction should not change if certain transformations are applied to the input data. For example, rotating a crystal structure or using a supercell of the same crystal structure should not affect a model's prediction of its formation energy per atom.

Recently, there have been several studies aimed at engineering neural network with desirable invariance behavior built in. Message passing neural networks [66] and the related graph neural networks [67, 68] are a new class of such methods that are quickly gaining popularity for materials property prediction. We discuss these models in more detail later in this chapter.

In other cases, properties should be equivariant, meaning they should transform the same as the input is transformed. Recent studies have also investigated equivariant neural networks that can predict properties of molecules and compounds with both magnitude and directionality, such as polarization or dielectric tensors, that maintain the underlying equivariance with respect to certain operations like translations and rotations. Tensor field networks [69] and E(n) equivariant graph neural networks [70] are examples of such methods.

5.11 Convolutional neural networks

The class of deep learning models responsible for the modern renaissance in image recognition and computer vision are convolutional neural networks, also known as CNNs. The goal behind CNNs is to create neural networks that can handle data distributions with certain equivariant/invariant features. For example, the location of a hand-written digit in an image should not change what digit a classifier should predict. In addition to normal fully connected hidden layers we have already discussed, CNNs models make use of specialized layers called *convolution layers* and *pooling layers*. Convolution layers apply learnable filters (also called kernels) to input tensors to produce feature maps as outputs. Pooling layers step the size of the input down to a smaller size using some sort of rule, such as dividing an input array into sub-arrays and creating an output array from the largest element of each sub-array (max pooling).

Because they are so adept at image processing, CNNs are most commonly used in the materials discovery space for processing microscopy [71], X-ray diffraction [72, 73], and spectroscopy data [74]. However, CNNs are also used for prediction properties from volumetric 3D data such as electron densities and atomic position grids. For example, Zhao et al. trained CNNs on volumetric electron charge density data and the Magpie descriptors of Ward et al. to predict the bulk and shear moduli of inorganic crystals [75].

Convolutional Neural Network

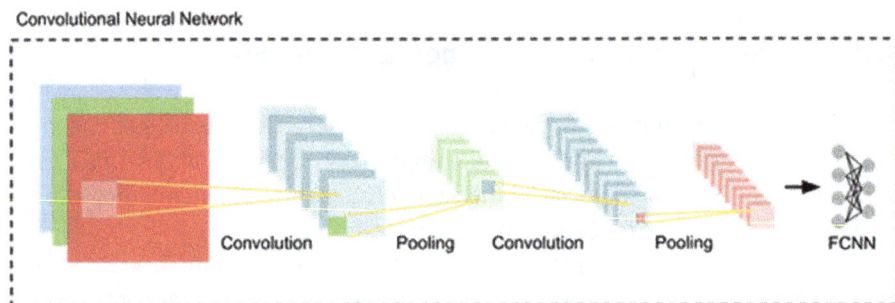

Fig. 5.5: Schematic of a convolutional neural network. Alternating convolution and pooling layers are applied before the features are passed to a fully connected neural network (FCNN). In real CNNs, there may be multiple "blocks" of convolutions, pooling, and FCNNs applied before a final output layer.

5.11.1 ResNets

Currently, the most common form of CNN is by far the residual network, aka ResNet [76]. The core breakthrough of ResNet was the *residual block* (Fig. 5.6), which

Residual Block

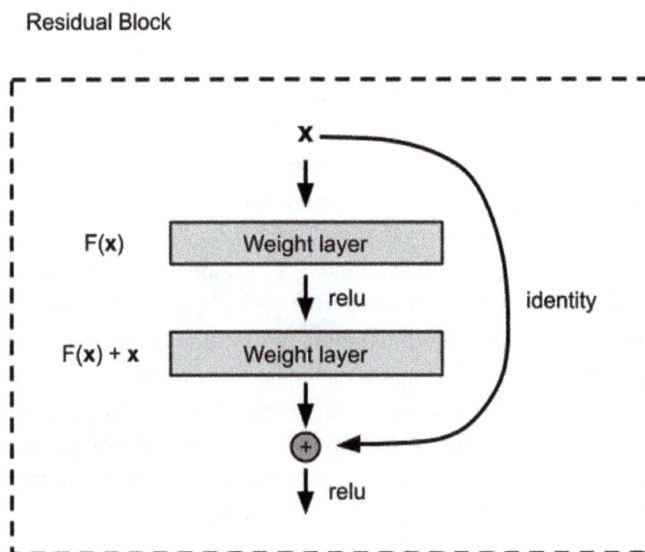

Fig. 5.6: Residual block diagram from the original Resnet paper [76]. Features are added to the output of later layers, a form of skip-connection.

helped overcome some of the problems that had traditionally limited the performance of very deep convolutional neural networks. However, the use of neural networks utilizing residual blocks is not restricted to CNNs. ResNets have been used for a wide variety of different deep learning tasks. One example from the materials space is the general purpose property prediction model IRNet by Jha et al. [77].

5.12 Graph neural networks

In 2007, Behler and Parrinello showed that the total potential energy of a system could be decomposed into effective atomic contributions with neural network potentials. Since then, Behler–Parrinello neural networks (BPNNs) [78] have become well-established for accelerating materials discovery, especially as a way to accelerate molecular dynamics simulations [79–81].

Graph neural networks (GNNs) are a newer class of machine learning algorithm that also analyze materials through the lens of local chemical environments around atoms. GNNs have shown state of the art performance on many property prediction tasks such as predicting band gaps and formation energies with high accuracy [33, 34, 82]. GNNs operate on graph-structured data, such as the network of bonds in a molecule or compound. This has the desirable property of being invariant to the orientation of the structure representation and also affords the user to directly use crystal structures as inputs instead of manually calculating and hand-selecting features, which has traditionally been the standard representation approach.

GNNs work by formulating the problem of predicting the properties of a graph as a sequential series of predictions for node, edge features, and global features. One version of this process is called "message passing," where the feature vectors from neighboring nodes in the graph are passed to a node and its state is updated with a neural network based on those features and the node's own internal state. In other variants of GNNs, the edges (bonds) of the graph have their own features and the overall graph itself also has an associated feature vector.

5.12.1 Crystal graph convolutional neural networks

Crystal graph convolutional neural networks (CGCNN) [33] are also an established graph-learning based approach to predict the properties of materials. CGCNN represents crystal structures as "crystal graphs" that encode atomic information, bonding interactions, and periodicity, and then builds a convolutional neural network on top of that graph to automatically extract representations for property prediction. The CNN built on top of the crystal graph is similar to conventional CNN architectures (i.e. convolution and pooling layers). However, in CGCNN, the convolution operation is a *graph convolution* that convolutes node features with features of neighboring nodes by concatenating node and edge features together, applying a linear transformations, and applying nonlinear activation functions to the result. The full form of how this convolution function updates the node features of node i at step t, $v_i^{(t)}$, is

$$v_i^{(t+1)} = v_i^{(t)} + \sum_{j,k} \sigma\left(z_{(i,j)_k}^{(t)} W_f^{(t)} + b_f^{(t)}\right) \odot g\left(z_{(i,j)_k}^{(t)} W_s^{(t)} + b_s^{(t)}\right)$$

where $W_f^{(t)}$ and $W_s^{(t)}$ are weight matrices, $b_f^{(t)}$ and $b_s^{(t)}$ are biases, and $z_{(i,j)}^{(t)}k$ is the concatenation of $v_i^{(t)}$ and the feature vector of the kth edge between nodes i and j, $u_{(ij)}^{(t)}k$. The functions σ and g are the sigmoid activation function and another activation function, respectively.

With each subsequent graph convolution layer applied to the node features, information from nodes that are increasingly farther away in the graph is integrated into the updated features. These features can then be optionally processed by a neural network before being pooled together to produce a feature vector for the graph as a whole (Fig. 5.7.).

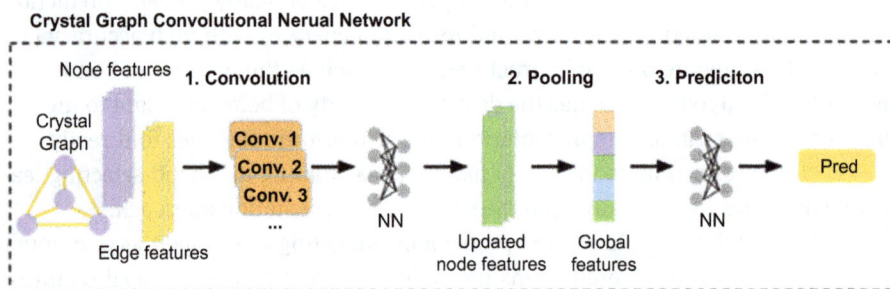

Fig. 5.7: Overview of crystal graph convolutional neural network algorithm.

5.12.2 MatErials graph network

MatErials graph network (MEGnet) [34] is slightly more modern than CGCNN in terms of the graph neural network algorithm behind its "MEGNet block." GNNs work by iteratively updating edge, node, and global features of a graph using neural networks in particular sequences. In the case of MEGNet, the sequence is edges ‒→nodes ‒→global.

At each time step t, the edge features $\mathbf{e}_k^t$ are updated by concatenating together ($\oplus$) the node features of the nodes it connects and the current edge features.

$$\mathbf{e}_k' = \phi_e\left(v_{s_k} \oplus v_{r_k} \oplus e_k \oplus \mathbf{u}\right)$$

This concatenated vector is passed to an update function, ϕ_e, (neural network) and new edge features e'_k are returned. Then, new node features v'_i for each node in the graph are calculated by collecting the features of every edge the node participates in, concatenating their average with the node's own embedding and the global feature vector, and passing that combined vector to an update function (NN) ϕ_v

$$\bar{v}_i^e = \frac{1}{N_i^e} \sum_{k=1}^{N_i^e} \{e'_k\}_{r_k=i}$$

$$v'_i = \phi_v\left(\bar{v}_i^e \oplus v_i \oplus u\right)$$

Finally, the global state of the graph is updated using information from the edges, nodes, and current global state as

$$\bar{v}_i^e = \frac{1}{N_i^e} \sum_{k=1}^{N_i^e} \{e'_k\}_{r_k=i}$$

$$v'_i = \phi_v\left(\bar{v}_i^e \oplus v_i \oplus u\right)$$

$$u' = \phi_u\left(\bar{u}^e \oplus \bar{u}^v \oplus u\right)$$

where ϕ_u is the global state update function (also a NN).

MEGNet has shown state of the art performance on a variety of materials property prediction tasks such as total energy (MAE $\leq$ 10 mev/atom).

5.13 Generative adversarial networks

For tasks where *new* structures or synthetic training data are desired, generative adversarial networks (GANs) can be used to generate new data points from the same overall distribution as a given dataset [83]. GANs are very elegant solutions to this problem in which two models are trained in competition with one another like an art forger and a detective tasked with identifying forgeries. The *generator* model of a GAN is trained to synthesize a new pieces of data; its loss function is based on whether it can fool a *discriminator* model, which is in turn trained to detect whether an input was genuine or synthesized by the generator. By training these two models together, the performance of both models improves until the generator is capable of producing very realistic synthetic data. Interested readers can see an example of GANs applied to human faces at the website https://thispersondoesnotexist.com, which randomly generates a new face each time the page is refreshed.

In materials discovery, GANs have been used most widely as a way to generate new candidate compositions and structures. For example, Xin et al. used the composition-generating MATGAN [85] in concert with an active learning process to generate and screen candidate compositions for new wide-band-gap materials [86]. Figure 5.8 shows crystal structures generated for Mg–Mn–O compositions using a GAN trained on structures from the Materials Project by Kim et al. [84].

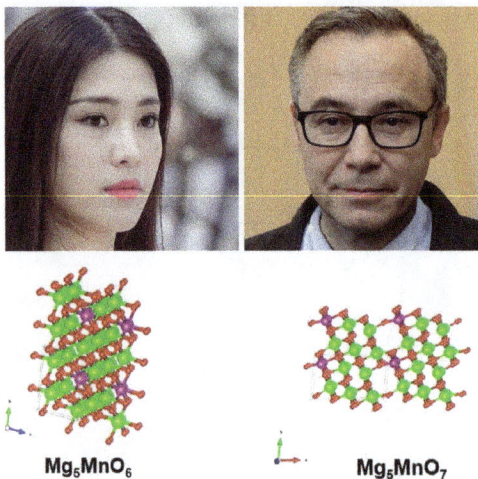

Mg_5MnO_6 Mg_5MnO_7

Fig. 5.8: (Top row) Images of faces generated with a GAN. (Bottom row) Crystal structures generated with a GAN [84].

5.14 Natural Language processing for accelerated materials discovery

The main barrier to using machine learning for materials discovery is not algorithms but rather access to high-quality training data. Unfortunately, the vast majority of human knowledge about materials exists in forms that are not readily accessible to these machine learning techniques: text, tables, and figures. Because valuable materials-property data are scattered across research papers in heterogeneous forms, we can only utilize a small fraction of this data to train our models today. In the last few years, however, there has been growing interest in using techniques from the field of natural language processing (NLP) to build structured databases of materials property data extracted from the literature and use unsupervised NLP techniques to learn directly on unstructured materials text.

5.15 Unsupervised learning

Unsupervised learning is the branch of machine learning concerned with learning from data that does not have associated labels whereas in the *supervised learning* examples we have seen up to this point all of our data have been paired with labels. The idea behind unsupervised learning is that there exists structure within data itself that we can leverage to learn useful patterns. Unsupervised learning has exploded in the last few years because it allows for researchers to greatly boost performance on supervised tasks by first pre-training models on large datasets of unlabeled data and then fine-tuning them on labeled sets. This has had an especially large impact on natural language processing.

5.15.1 Word2vec

Almost all machine learning models require numerical vector inputs which poses a problem for training ML models on human language. How do we make feature vectors out of words? The naive way is to create an N-dimensional basis out of our vocabulary of N words where each word vector is represented as the unit vector for one of the N directions. This is called one-hot encoding.

As one might expect, one-hot encoding is not a particularly efficient or effective description for word meaning, and one of the main hurdles to overcome in the field of NLP was somehow creating rich, low-dimensional, feature vectors for words that encode their meaning and relationship to other words. In 2013, researchers at Google made a giant leap forward in word embedding (constructing word vectors) and natural language processing with an algorithm known as Word2Vec. Word2vec leverages the insight that similar words appear in similar contexts in text. For example, "dog" and "cat" both represent furry, four-legged animals that are kept as pets and these words both frequently co-occur with words like "fur," "paw," "feed," "house," etc. Accordingly, words with disparate meanings are less likely to occur in similar contexts. The Word2vec authors proposed learning word embeddings by structuring the problem as a co-occurrence prediction task with a very simple neural network. Word2vec uses a neural network with a single hidden layer (linear activation) to predict whether words (encoded as one-hot vectors, see Fig. 5.9) co-occur within a certain window size, usually about eight words. By training this architecture to perform this task on millions of passages of text, the meanings of words become encoded in the weights of the neural network and individual word vectors can be extracted from the learned weight matrix.

The word embeddings of Word2vec not only greatly improved the performance of downstream natural language processing models that used the word embeddings as inputs, but they also had interesting intrinsic features such as latent directions in the embedding space that map onto intuitive concepts. Mathematically, these

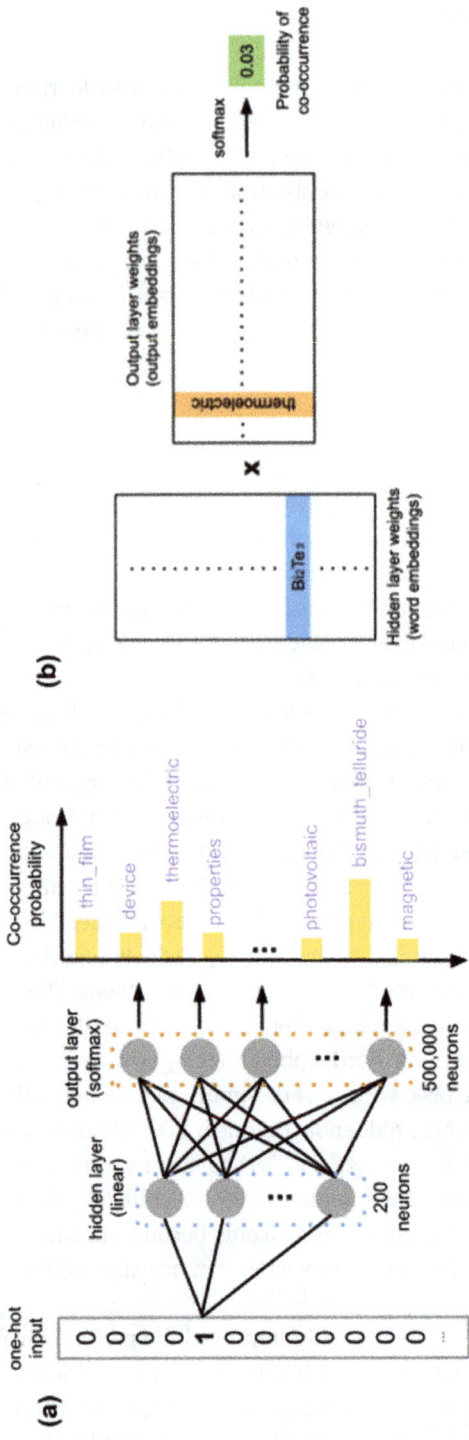

Fig. 5.9: (a) Overview of the word2vec algorithm for materials discovery. Word embeddings are derived by training a neural network to predict word co-occurrence. (b) Word embeddings are derived from the weights of the neural network (red.) Note: This is a placeholder figure unless we can get rights from nature.

directions can be constructed with vector subtraction and addition, i.e. "king" – "man" + "woman" = ?. The vectors sum of these word embeddings in this expression results in a new vector that is closer to the word embedding for "queen" than that of any other word in the vocabulary.

5.15.2 Word2Vec for materials discovery

Naturally, one might wonder what Word2vec might learn when trained on materials science literature. This was the premise of Tshitoyan et al., which we co-authored. In this study, we trained word2vec on 1.5 million materials science abstracts to generate the Mat2vec word embedding dataset [87].

In fact, these models learn word embeddings that are rich in concepts relevant to chemistry and materials properties such as elemental properties and the structure of the periodic table (Fig. 5.11). Remarkably, the vector addition/subtraction operations of Word2Vec also maps on to analogies in the materials domain, for instance magnetic ordering:

$$\text{"ferromagnetic} - NiFe + IrMn \approx \text{antiferromagnetic"}$$

Perhaps more surprisingly, unsupervised word embeddings can be directly utilized to discover new functional materials and uncover relationships that may not be readily apparent to human researchers who have only read a subset of the available literature. One can screen the vocabulary for instances of material names that have high similarity to a desired application but do not co-occur with it in (or related words) in any passages. In our study, we put this method into practice and showed that unsupervised word embeddings derived from the abstracts of past publications could be used to predict future thermoelectric discoveries. This pattern also appears to be the case for other types of materials including photovoltaics, ferroelectrics, and topological insulators (Fig. 5.10.)

We also have observed that the word embeddings of for the names of the elements learned by the unsupervised model encode information relevant to atomic properties of the elements such as covalent radius and electronegativity. To show this, we performed a PCA on the element name embeddings and then trained linear regressions for a number of properties (Fig. 5.11.)

5.15.3 Data extraction and insights from NLP

While data extraction from text is not directly used for materials discovery, the large datasets of materials property data that will be required by the next generation of models will likely be derived through text mining.

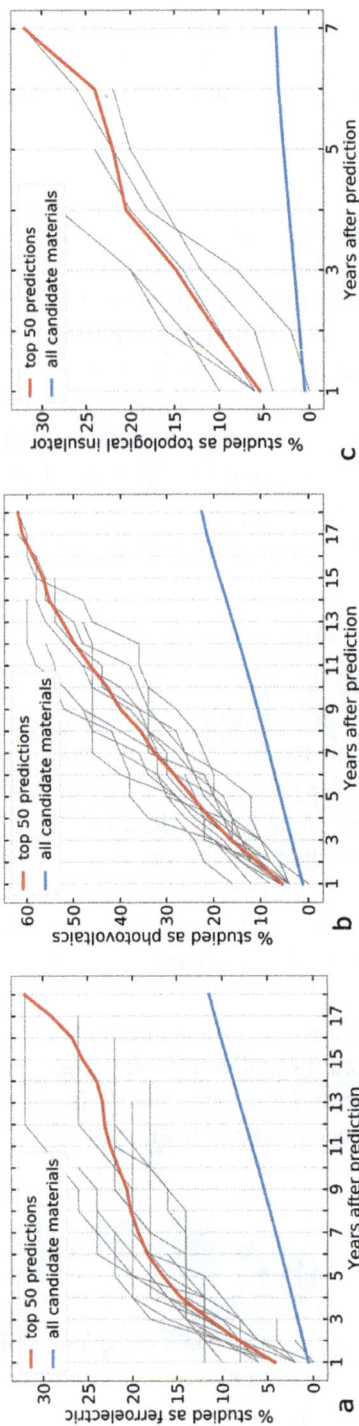

Fig. 5.10: Probability that word2vec predictions are confirmed as a function n years after prediction was made for: (a) ferroelectrics, (b) photovoltaics, and (c) topological insulators. Word2vec (red line) is 3–6x better than random guessing (blue line) despite receiving no explicit information about materials chemistry.

(a)

(b)

(c)

Property	R^2
Mendeleev number	74.4
Atomic weight	72.4
Melting T	73.9
Covalent radius	81.8
Electronegativity	76.2
Row	80.7
Column	69.2

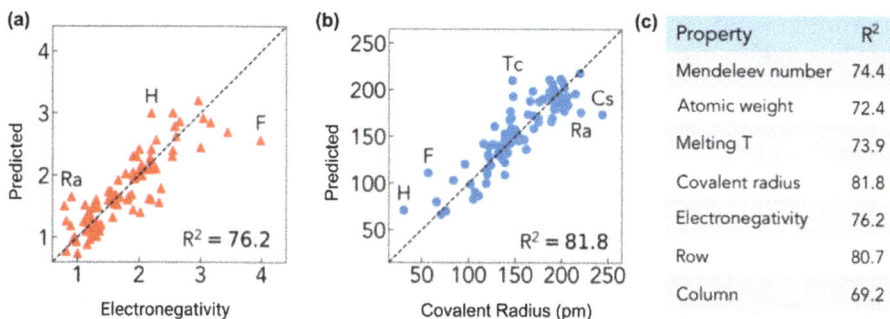

Fig. 5.11: Predictions of elemental properties from word embeddings of element names. a-b. Electronegativity and covalent radius of elements predicted using linear regression with cross-validation. The first 15 principal components of word embeddings of element names (e.g. "hydrogen") were used as features. c. Mean R2 score (in percent) from 20-fold cross validation.

When paired with pretrained word embeddings, neural networks can be used to extract useful data out of materials science text passages. One of the most common data extraction tasks is named entity recognition (NER), in which words in passages are labeled with desired tags such as "material" or "property." In this way, large amounts of structured materials data (such as material-property co-occurrence statistics) can be extracted out of unstructured text passages.

5.15.4 Transformer-based models (BERT and GPT-3)

Word2Vec is not the only word embedding method. Other popular word embedding techniques, such as GloVe [88], can produce similar results. More recently, however, new methods for unsupervised learning that learn context-sensitive word embeddings have been developed that significantly outperform GloVe and Word2Vec on downstream tasks. The key advancement behind these methods is a mechanism called "attention" and the "transformer" architecture [89] that utilizes it.

Attention mimics cognitive attention by applying context-sensitive weighting to different sections of the input. When multiple attention "heads" are used, it results in "multi-headed attention," which runs several attention mechanisms in parallel. Transformers make extensive use of multi-headed attention to achieve their expressive powers. While this gives these models a more complex architecture that requires orders of magnitude more data to train, it produces rich word embeddings and these models can perform astounding tasks such as writing passages that are nearly indistinguishable from human writing [90], generating functional code from a short description of a program [91], and even linking images and text together into models capable of few-shot image classification [74] or generating realistic images from text descriptions alone [92]. The two most well-known types of transformers for natural language processing are BERT-based models [93, 94] and GPT [90, 95, 96].

Because these models are so large (billions of parameters), they are typically trained once on massive general-purpose corpora (hundreds of billions of words) and then fine-tuned on smaller task-specific data sets. BERT and GPT-based models trained pretrained on scientific literature exhibit greatly improved performance on domain-specific NLP tasks [97] including materials NER. However, there remains a large amount of work to be done in this space before we can reliably extract all of the relevant materials data, namely models for linking entities together into relational databases.

5.15.5 Other methods for data extraction

Large pretrained language models paired with supervised fine-tuning is not the only way to do effective data extraction from unstructured text. The ChemDataExtractor project out of Cambridge has shown strong performance on both NER and entity linking for creating structured datasets and utilizes more rule-based methods over transformer-based language models [98, 99].

5.15.6 Accelerating synthesis of new materials through text mining

In order to truly be discovered, a material must eventually be synthesized in the laboratory and experimentally tested. Unfortunately, synthesis is often the most difficult and time-consuming stage of materials discovery. To this day, it remains mostly a trial-and-error process and it can take many years for predicted materials to be realized in the lab. However, the predictable nature of well-understood chemical reactions and materials processing techniques also implies that definitive rules and procedures for the synthesis of nearly any realizable compound should, in theory, be possible to predict. Machine learning has great potential to help bridge this frustrating gap between what we know should be possible and what we can make in the lab [100].

Because so much of the materials research community's knowledge about synthesis has been expressed in the methods sections of research papers, there has been a growing interest in using natural language processing techniques to extract synthesis data from the literature and use it to train machine learning models that can aid researchers in designing synthesis routes for new materials. To date, most work in synthesis text mining has been restricted to the fundamentals that will need to be in place before full datasets of high-quality synthesis graphs (full descriptions of how all synthesis steps and conditions are related) can be mined from the literature. These include extracting and classifying synthesis recipes into categories, such as differentiating between solid state and hydrothermal synthesis [101], extracting synthesis parameters for oxide systems [102], and building ontologies for synthesis knowledge [103]. However, work is now beginning to appear where full entity and relation data for synthesis reactions is automatically extracted from unstructured text. For example, Kuniyoshi et al. have demonstrated a system for extracting synthesis graphs for solid-state battery compounds using a combination of neural network based NER and rule-based relationship extraction [104].

Recent work has also shown that text mining synthesis recipes from the methods sections of research papers is a promising route toward better understanding of materials synthesis. In one study, Kim et al. used an autoencoder trained on synthesis routes extracted from the literature for $SrTiO_3$, brookite TiO_2, and MnO_2. They found that the latent space could be used to explore the driving factors behind formation of various phases, such as the role of certain ions and precursors in polymorph selectivity [105]. In another study by He et al., the researchers used named entity recognition to collect a dataset of the targets and precursors of solid-state synthesis reactions and then analyzed the target-precursor co-occurrence statistics in this dataset to build a probabilistic models for precursor substitution and similarity [106].

The quick pace of advances in this field will likely lead to large datasets of high-quality synthesis information becoming available within the next 5 to 10 years along with models that can help researchers gain better understanding of how various conditions, precursors, and processing techniques can be used to tune synthesis pathways for novel compounds.

5.16 Dimensionality reduction

There is a general principle in machine learning that the number of parameters required by a model to fit a dataset increases with the size of the data's dimension. That is, in general models need more parameters to fit data that is 10-dimensional than data that is two-dimensional. Conversely, there is also a principle that larger models require more data to train because large models trained on few data points tend to over-fit. Luckily, in many cases, it is possible to reduce the number of features while still retaining most of the information in the original data points through a process known as *dimensionality reduction*.

5.16.1 Principal component analysis and nonnegative matrix factorization (NMF)

Principle component analysis (PCA) is the best-known and most widely used algorithm for dimensionality reduction. PCA constructs a projection that transforms N dimensional data into M dimensional data (where M < N) that also maximizes the variance of the final data. This effectively maximizes the amount of information retained in the new, smaller data points from the original high-dimensional data. PCA is well covered in many other machine learning resources and efficient implementations are available in many python libraries, such as scikit-learn.

In a brilliant application of PCA, Curtarolo et al. showed that applying PCA to total energy calculations for compounds revealed that energies for 114 structure prototypes could be inferred from energies for a subset of them. They then showed that a small number of calculations could be done for certain structure prototypes, which could then be used to infer the energy at that composition for a larger number (114) of structure prototypes and infer the stable structure prototype without exhaustively testing all of them [107].

A related technique is nonnegative matrix factorization, which is similar to PCA but also adds the constraint that data points are composed of strictly additive components. NMF has been used for automated XRD pattern analysis and phase identification [22].

5.16.2 Other methods for dimensionality reduction

The best way to visualize data that is distributed in spaces with more than three dimensions has been an open research question for many years. While PCA can be used for this task, it does not reproduce the local neighborhood of data points well. T-stochastic neighbor embeddings (TSNE)[108] is a widely-used method for visualizing high dimensional data. It attempts to embed a dataset into a low (2 or 3) dimensional space in such a way that points that were neighbors in the high dimensional space are neighbors in the low dimensional space. However, it has recently fallen out of favor due to some issues that have since been corrected by newer methods. The main flaw of TSNE is that does not preserve the global structure of the data well. Within clusters, distances between points is approximately preserved, but this is not so between clusters or between points that are not close together in the original dataset. There are also performance reasons that TSNE is now disfavored and newer methods scale to higher dimensions (beyond 2–3) and larger datasets much better. Uniform manifold approximation and projection (UMAP) is one of these updated methods. UMAP preserves more global structure in the dataset than TSNE without sacrificing local structure and it can be used on large datasets and scales to dimensions beyond 2–3 more efficiently than TSNE. Figure 5.12 shows the UMAP and TSNE embeddings for the Mat2vec word embeddings from Tshitoyan et al. [87].

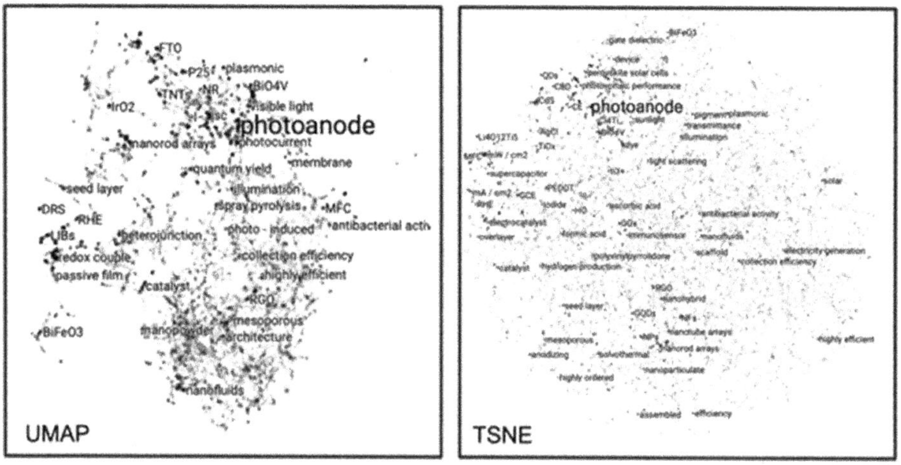

Fig. 5.12: UMAP and TSNE projections of the 200-dimensional Mat2vec embeddings with the top 100 words similar to "photoanode" highlighted in blue. Darker blue points are closer to "photoanode" in the original 200-D vector space than lighter blue points. Note how UMAP does a better job of retaining near-neighbors (blue points) in close proximity to each other.

5.16.3 Autoencoders

Autoencoders are unsupervised machine learning models that learn a reversible transformation between a high dimensional vector space (such as the space of all images) and a compressed, lower dimensional *latent space.* Autoencoders are made of two parts, an encoder and a decoder, which are trained in tandem by defining a "reconstruction" loss that compares the original input to the output of the model's decoder. An input passes through the encoder and is embedded in a "latent space" that is a lower dimension than the original input. The latent representation is then passed to the decoder and re-expanded back to the original dimension before the reconstruction loss is calculated against the original input. The benefit of constructing latent features is that models trained on them typically need fewer parameters and potentially exhibit less over-fitting. Machine learning models can also be trained on the latent representations in order to explore the latent space for new structures that may have improved properties. Candidate points with good predictions can be identified and then passed to the decoder to generate (hopefully) valid structures with improved properties.

An example of this process is the iMatGen model, which is an autoencoder developed Noh et al. that was used to design new solid-state V_xO_y materials [109].

5.17 Opportunities in ML for materials discovery

Despite so much progress in the last 10 years using machine learning for materials discovery, we have just scratched the surface of what is possible. The next decade will be defined by innovative programs that combine computational modeling, machine learning, AI-assisted synthesis, and automated characterization [100]. Our view is that there are a number of especially fertile areas available

Solving data scarcity. By far and away, the greatest barrier of modern ML for materials is the scarcity of data. The efficacy of machine learning relies entirely on data; without data – and in particular, structured data – machine learning's inferential power is severely handicapped. While online repositories of DFT calculations have made datasets of a select few solid-state *ab initio* properties more abundant, the vast majority of existing experimental property data remains in forms that are inaccessible to our machine learning models. However, property data in literature presents an incredible opportunity for data mining. Hidden inside the text and tables of millions of scientific publications lies materials property data which has never been collated into structured collections. Using this vast corpus of data in combination with natural language processing, in particular models for named entity recognition and entity relationship modeling, is a tractable avenue to producing structured ML datasets unprecedented in size and scope.

Transformers for structure prediction. Predicting stable crystal structures from compositions remains a grand challenge of materials discovery. Current approaches like ab initio random structure search [110] and USPEX [111] still require a large number of structure relaxations for every composition. Advancements in parallel fields, such as the remarkable performance of the attention based AlphaFold network [112] for protein structure prediction, are a promising sign that similar advancements can be made in the domain of materials. Composition-only attention-based models such as RooSt [35] and CRABNet [36] (which predict properties directly from composition) are also indications that attention networks can adequately model complex materials information. Yet, the advances in attention-based deep neural networks have not been extensively investigated for diverse crystal structure prediction to date. We believe that this represents a very promising opportunity for the application of transformers.

ML predictions for device-level performance. There remains a large gap between predicting bulk properties for materials and the actual in-device performance that can be expected. This is because materials exhibit complex multi-scale behavior and the processing conditions and operating environments, they are exposed to can introduce drastic changes to their properties. Right now, materials ML is in its infancy, but some day in the future it may be possible to use machine learning to predict some of these complex interactions.

References

[1]   Davies DW et al. Computational screening of all stoichiometric inorganic materials. Chem, 2016, 1, 617–627. issn: 24519294.

[2]   Council NR. Materials in the New Millennium: Responding to Society's Needs Isbn: 978-0-309-07562-6. The National Academies Press, Washington, DC, 2001, doi: 10.17226/10187. <https://www.nap.edu/catalog/10187/materials-in-the-new-millennium-responding-to-societys-needs>.

[3]   Villars P, Cenzual K. Pearson's Crystal Data: Crystal Structure Database for Inorganic Compounds (on DVD) (ASM International®, 2021).

[4]   Villars P et al. The Pauling File in European Powder Diffraction EPDIC 8 443. Trans Tech Publications Ltd, Jan. 2004), 357–360. doi: 10.4028/www.scientific.net/MSF.443-444.357.

[5]   Saal JE, Kirklin S, Aykol M, Meredig B, Wolverton C. Materials design and discovery with high-throughput density functional theory: the open quantum materials database (OQMD). Jom, 2013, 65, 1501–1509. issn: 10474838.

[6]   Kirklin S et al. The open quantum materials database (OQMD): Assessing the accuracy of DFT formation energies. Npj Comput Mater, 2015, 1, 15010.

[7]   Jain A et al. The Materials Project: A materials genome approach to accelerating materials innovation. APL Mater, 2013, 1, 011002. issn: 2166532X.

[8]   Ong SP et al. The materials application programming interface (API): A simple, flexible and efficient API for materials data based on Representational State Transfer (REST) principles. Comput Mater Sci, 2015, 97, 209–215.

[9]   Curtarolo S et al. AFLOW: An automatic framework for high-throughput materials discovery. Comput Mater Sci, 2012, 58, 218–226. issn: 09270256.

[10]   De Jong M et al. Charting the complete elastic properties of inorganic crystalline compounds. Sci Data, 2015, 2, doi: 10.1038/sdata.2015.9.

[11]   Meredig B et al. Combinatorial screening for new materials in unconstrained composition space with machine learning. Phys Rev B Condens Matter Mater Phys, 2014, 89, issn: 10980121. doi: 10.1103/PhysRevB.89.094104.

[12]   Tehrani AM et al. Machine learning directed search for ultraincompressible, superhard materials. J Am Chem Soc, 2018, 140, 9844–9853.

[13]   Masood H, Toe CY, Teoh WY, Sethu V, Amal R. Machine learning for accelerated discovery of solar photocatalysts. ACS Catal, 2019, 9, 11774–11787.

[14]   Ricci F, Dunn A, Jain A, Rignanese GM, Hautier G. Gapped metals as thermoelectric materials revealed by high-throughput screening. J Mater Chem A, 2020, 8, 17579–17594. issn: 20507496.

[15]   Ward L et al. A machine learning approach for engineering bulk metallic glass alloys. Acta Mater, 2018, 159, 102–111.

[16]   Sun W et al. Machine learning–assisted molecular design and efficiency prediction for high-performance organic photovoltaic materials. Sci Adv, 2019, 5. doi: 10. 1126/sciadv.aay4275. https://doi.org/10.1126/sciadv.aay4275.

[17]   Lu S et al. Accelerated discovery of stable lead-free hybrid organic-inorganic perovskites via machine learning. Nat Commun, 2018, 9. doi: 10.1038/s41467018-05761-w. https://doi.org/10.1038/s41467-018-05761-w.

[18]   Frey NC et al. High-throughput search for magnetic and topological order in transition metal oxides. Sci Adv, 2020, 6. doi: 10.1126/sciadv.abd1076. https://doi.org/10.1126/sciadv.abd1076.

[19]   Shandiz MA, Gauvin R. Application of machine learning methods for the prediction of crystal system of cathode materials in lithium-ion batteries. Comput Mater Sci, 2016, 117, 270–278.

[20] Balachandran PV, Kowalski B, Sehirlioglu A, Lookman T. Experimental search for high-temperature ferroelectric perovskites guided by two-step machine learning. Nat Commun, 2018, 9. doi: 10.1038/s41467-018-03821-9. <https://doi.org/10.1038/s41467-018-03821-9>.

[21] Domínguez LA, Goodall R, Todd I. Prediction and validation of quaternary high entropy alloys using statistical approaches. Mater Sci Technol, 2015, 31, 1201–1206.

[22] Stanev V et al. Unsupervised phase mapping of X-ray diffraction data by nonnegative matrix factorization integrated with custom clustering. Npj Comput Mater, 2018, 4, issn: 20573960. doi: 10.1038/s41524-018-0099-2. arXiv: 1802.07307. http://dx.doi.org/10.1038/s41524-018-0099-2.

[23] Schleder GR, Padilha AC, Acosta CM, Costa M, Fazzio A. From DFT to machine learning: recent approaches to materials science – A review. JPhys Mater, 2019, 2. issn: 25157639. doi: 10.1088/2515-7639/ab084b.

[24] Vasudevan RK et al. Materials science in the artificial intelligence age: highthroughput library generation, machine learning, and a pathway from correlations to the underpinning physics. MRS Commun, 2019, 9, 821–838. issn: 21596867.

[25] Butler KT, Davies DW, Cartwright H, Isayev O, Walsh A. Machine learning for molecular and materials science. Nature, 2018, 559, 547–555. issn: 14764687.

[26] Schmidt J, Marques MR, Botti S, Marques MA. Recent advances and applications of machine learning in solid-state materials science. Npj Comput Mater, 2019, 5, issn: 20573960. doi: 10.1038/s41524-019-0221-0. <http://dx.doi.org/10.1038/s41524-019-0221-0>.

[27] Morgan D, Jacobs R. Opportunities and challenges for machine learning in materials science. AnnuRev Mater Res, 2020, 50, 71–103. issn: 15317331.

[28] Hill R The elastic behaviour of a crystalline aggregate. Proceedings of the Physical Society. Section A, 65, 349–354, May 1952.

[29] Oliynyk AO et al. High-throughput machine-learning-driven synthesis of fullHeusler compounds. Chem Mater, 2016, 28, 7324–7331.

[30] Ward L, Agrawal A, Choudhary A, Wolverton C. A general-purpose machine learning framework for predicting properties of inorganic materials. Npj Comput Mater, 2016, 2, 1–7. issn: 20573960.

[31] Goldschmidt VM. Die Gesetze der Krystallochemie. Die Naturwissenschaften, 1926, 14, 477–485.

[32] Jha D et al. ElemNet: deep learning the chemistry of materials from only elemental composition. Sci Rep, 2018, 8, 1–13. issn: 20452322.

[33] Xie T, Grossman JC. Crystal graph convolutional neural networks for an accurate and interpretable prediction of material properties. Phys Rev Lett, 2018, 120, 145301. issn: 10797114.

[34] Chen C, Ye W, Zuo Y, Zheng C, Ong SP. Graph networks as a universal machine learning framework for molecules and crystals. Chem Mater, 2019, 31, 3564–3572. issn: 15205002.

[35] Goodall RE, Lee AA. Predicting materials properties without crystal structure: Deep representation learning from stoichiometry. Nat Commun, 2020, 11, 1–9. issn: 20411723.

[36] Wang AYT, Kauwe SK, Murdock RJ, Sparks TD. Compositionally restricted attention-based network for materials property predictions. Npj Comput Mater, 2021, 7, 1–10. issn: 20573960.

[37] Dunn A, Wang Q, Ganose A, Dopp D, Jain A. Benchmarking materials property prediction methods: The matbench test set and automatminer reference algorithm. Npj Comput Mater, 2020, 6. doi: 10.1038/s41524-020-00406-3. https://doi.org/10.1038/s41524-020-00406-3.

[38] Hastie T, Tibshirani R, Friedman J. The Elements of Statistical Learning, New York, NY, USA, Springer New York Inc., 2001.

[39] Murphy KP. Machine Learning A Probabilistic Perspective. MIT Press, Cambridge, Massachusetts, 2012, 429–439. isbn: 978-0262-01802-9. doi: 10.1111/j.1468-0394.1988. tb00341.x.

[40] Ren F et al. Accelerated discovery of metallic glasses through iteration of machine learning and high-throughput experiments. Sci Adv, 2018, 4, issn: 23752548. doi: 10.1126/sciadv. aaq1566.

[41] Min K, Choi B, Park K, Cho E. Machine learning assisted optimization of electrochemical properties for Ni-rich cathode materials. Sci Rep, 2018, 8, 1–7. issn: 20452322.

[42] Breiman L. Bagging predictors. Mach Learn, 1996, 24, 123–140. issn: 08856125.

[43] Breiman L. Random forests. Mach Learn, 2001, 45, 5–32.

[44] Pedregosa F et al. Scikit-learn: machine learning in python. J Mach Learn Res, 2011, 12, 2825–2830. issn: 1532-4435.

[45] Chen T, Guestrin C. XGBoost: A scalable tree boosting system. arXiv. arXiv: 1603.02754v3, 2016.

[46] Schleder GR, Focassio B, Fazzio A. Machine learning for materials discovery: two-dimensional topological insulators. Appl Phys Rev, 2021, 8, issn: 19319401. arXiv: 2107.07028. doi: 10.1063/5.0055035.

[47] Bottou L, Curtis FE, Nocedal J. Optimization methods for large-scale machine learning. SIAM Review, 2018, 60, 223–311. issn: 00361445.

[48] Nesterov Y In Introductory Lectures on Convex Optimization: A Basic Course. Springer US, Boston, MA, 2004, 51–110. isbn: 978-1-4419-8853-9. doi: 10.1007/9781-4419-8853-9_2. <https://doi.org/10.1007/978-1-4419-8853-9_2>.

[49] Kingma DP, Ba J. Adam: a method for stochastic optimization, 2017. arXiv: 1412.6980[cs.LG].

[50] Talapatra A et al. Autonomous efficient experiment design for materials discovery with Bayesian model averaging. Phys Rev Mater, 2018, 2. doi: 10.1103/physrevmaterials.2.113803. <https://doi.org/10.1103/physrevmaterials.2.113803>.

[51] Gopakumar AM, Balachandran PV, Xue D, Gubernatis JE, Lookman T. Multi-objective optimization for materials discovery via adaptive design. Sci Rep, 2018, 8. doi: 10.1038/ s41598-018-21936-3. <https://doi.org/10.1038/s41598-018-21936-3>.

[52] Lourenço MP, Dos Santos Anastácio A, Rosa AL, Frauenheim T, Da Silva MC. An adaptive design approach for defects distribution modeling in materials from first-principle calculations. J Mol Model, 2020, 26. doi: 10.1007/s00894-020-04438-w. https://doi.org/10. 1007/s00894-020-04438-w.

[53] Lourenço MP et al. A new active learning approach for global optimization of atomic clusters. Theor Chem Acc, 2021, 140, doi: 10.1007/s00214-02102766-5. <https://doi.org/10.1007/ s00214-021-02766-5>.

[54] Xue D et al. Accelerated search for materials with targeted properties by adaptive design. Nat Commun, 2016, 7. doi: 10.1038/ncomms11241. <https://doi.org/10.1038/ncomms11241>.

[55] Ueno T, Rhone TD, Hou Z, Mizoguchi T, Tsuda K. COMBO: an efficient Bayesian optimization library for materials science. Mater Discovery, 2016, 4, 18–21.

[56] Montoya JH et al. Autonomous intelligent agents for accelerated materials discovery. Chem Sci, 2020, 11, 8517–8532.

[57] Dunn A, Brenneck J, Jain A. Rocketsled: A software library for optimizing highthroughput computational searches. J Phys Mater, 2019, 2, 034002.

[58] Jones DR, Schonlau M, Welch WJ. Efficient global optimization of expensive black-box functions. J Global Opt, 1998, 13, 455–492. issn: 09255001.

[59] Rasmussen C. Gaussian processes for machine learning, MIT Press, Cambridge, Mass, 2006. isbn: 0-262-18253-X.

[60] Bucior BJ et al. Energy-based descriptors to rapidly predict hydrogen storage in metal-organic frameworks. Mol Syst Des Eng, 2019, 4, 162–174. issn: 20589689.

[61] Sendek AD et al. Machine learning-assisted discovery of solid Li-Ion conducting materials. Chem Mater, 2019, 31, 342–352. issn: 15205002.

[62] Sendek AD et al. A new solid Li-ion electrolyte from the crystalline lithiumBoron-sulfur system. SSRN Electron J, 2019, issn: 1556-5068. doi: 10.2139/ssrn.3404263.

[63] Mannodi-Kanakkithodi A, Pilania G, Huan TD, Lookman T, Ramprasad R. Machine learning strategy for accelerated design of polymer dielectrics. Sci Rep, 2016, 6, 1–10. issn: 20452322.

[64] Lecun Y, Bengio Y, Hinton G. Deep learning. Nature, 2015, 521, 436–444. issn: 14764687.

[65] Srivastava N, Hinton G, Krizhevsky A, Sutskever I, Salakhutdinov R. Dropout: a simple way to prevent neural networks from overfitting. J Mach Learn Res, 2014, 15, 1929–1958.

[66] Gilmer J, Schoenholz SS, Riley PF, Vinyals O, Dahl GE. Neural message passing for quantum chemistry. 34th International Conference on Machine Learning, ICML 2017, 3, 2053–2070, 2017.

[67] Scarselli F, Gori M, Tsoi AC, Hagenbuchner M, Monfardini G. The graph neural network model. IEEE Trans Neural Netw, 2009, 20, 61–80. issn: 10459227.

[68] Battaglia PW et al. Relational inductive biases, deep learning, and graph networks. arXiv, 1–40. issn: 23318422, 2018.

[69] Thomas N et al. Tensor field networks: Rotation- and translation-equivariant neural networks for 3D point clouds. issn: 2331-8422. arXiv: 1802.08219. <http://arxiv.org/abs/1802.08219>, 2018.

[70] Satorras VG, Hoogeboom E, Welling M. E(n) Equivariant Graph Neural Networks. arXiv: 2102.09944. <http://arxiv.org/abs/2102.09944>, 2021.

[71] Cang R, Li H, Yao H, Jiao Y, Ren Y. Improving direct physical properties prediction of heterogeneous materials from imaging data via convolutional neural network and a morphology-aware generative model. Comput Mater Sci, 2018, 150, 212–221. issn: 09270256.

[72] Park WB et al. Classification of crystal structure using a convolutional neural network. IUCrJ, 2017, 4, 486–494. issn: 20522525.

[73] Oviedo F et al. Fast and interpretable classification of small X-ray diffraction datasets using data augmentation and deep neural networks. Npj Comput Mater, 2019, 5, 1–9. issn: 20573960.

[74] Fukuhara M, Fujiwara K, Maruyama Y, Itoh H. Feature visualization of Raman spectrum analysis with deep convolutional neural network. Anal Chim Acta, 2019, 1087, 11–19. issn: 18734324.

[75] Zhao Y et al. Predicting elastic properties of materials from electronic charge density using 3D deep convolutional neural networks. J Phys Chem C, 2020, 124, 17262–17273. issn: 19327455.

[76] He K, Zhang X, Ren S, Sun J Deep residual learning for image recognition. Proceedings of the IEEE Computer Society Conference on Computer Vision and Pattern Recognition 2016-Decem, 770–778. issn: 10636919, 2016.

[77] Jha D et al. IRNet: A general purpose deep residual regression framework for materials discovery. Proceedings of the ACM SIGKDD International Conference on Knowledge Discovery and Data Mining, 2385–2393, 2019.

[78] Behler J, Parrinello M. Generalized neural-network representation of high-dimensional potential-energy surfaces. Phys Rev Lett, 2007, 98, 1–4. issn: 00319007.

[79] Khaliullin RZ, Eshet H, Kühne TD, Behler J, Parrinello M. Nucleation mechanism for the direct graphite-to-diamond phase transition. Nat Mater, 2011, 10, 693–697. issn: 14764660.

[80] Artrith N, Kolpak AM. Grand canonical molecular dynamics simulations of CuAu nanoalloys in thermal equilibrium using reactive ANN potentials. Comput Mater Sci, 2015, 110, 20–28. issn: 09270256.

[81] Hellström M, Quaranta V, Behler J. One-dimensional vs. two-dimensional proton transport processes at solid-liquid zinc-oxide-water interfaces. Chem Sci, 2019, 10, 1232–1243. issn: 20416539.

[82] Louis SY et al. Graph convolutional neural networks with global attention for improved materials property prediction. Phys Chem Chem Phys, 2020, 22, 18141–18148. issn: 14639076.

[83] Goodfellow I et al. Generative adversarial nets. Adva Neural Inf Process Syst, 2014, 27.

[84] Kim S, Noh J, Gu GH, Aspuru-Guzik A, Jung Y. Generative adversarial networks for crystal structure prediction. ACS Cent Sci, 2020, 6, 1412–1420. issn: 23747951.

[85] Dan Y et al. Generative adversarial networks (GAN) based efficient sampling of chemical composition space for inverse design of inorganic materials. Npj Comput Mater, 2020, 6, 1–7. issn: 20573960.

[86] Xin R et al. Active-learning-based generative design for the discovery of wideBand-gap materials. J Phys Chem C, 2021, 125, 16118–16128. issn: 19327455.

[87] Tshitoyan V et al. Unsupervised word embeddings capture latent knowledge from materials science literature. Nature, 2019, 571, 95–98. issn: 14764687.

[88] Pennington J, Socher R, Manning C. Glove: global vectors for word representation. Proceedings of the 2014 Conference on Empirical Methods in Natural Language Processing (EMNLP), 1532–1543. issn: 10495258, 2014.

[89] Vaswani A et al. Attention Is All You Need. 2017. arXiv: 1706.03762[cs.CL].

[90] Brown TB et al. Language models are few-shot learners. Advances in Neural Information Processing Systems 2020-December. issn: 10495258. arXiv: 2005. 14165, 2020.

[91] Chen M et al. Evaluating Large Language Models Trained on Code. arXiv: 2107. 03374. <http://arxiv.org/abs/2107.03374>, 2021.

[92] Ramesh A et al. Zero-Shot Text-to-Image Generation. 200. arXiv: 2102.12092. <http://arxiv.org/abs/2102.12092>, 2021.

[93] Devlin J, Chang M-W, Lee K, Toutanova K. BERT: Pre-training of Deep Bidirectional Transformers for Language Understanding. arXiv: 1810.04805. <http://arxiv.org/abs/1810.04805>, 2018.

[94] Liu Y et al. RoBERTa: A Robustly Optimized BERT Pretraining Approach. issn: 2331-8422. arXiv: 1907.11692. <http://arxiv.org/abs/1907.11692>, 2019.

[95] Radford A, Narasimhan K. Improving Language Understanding by Generative Pre-Training in, 2018.

[96] Radford A et al. Language Models are Unsupervised Multitask Learners. <https://d4mucfpk sywv.cloudfront.net/better-language-models/language-models.pdf>, 2018.

[97] Beltagy I, Lo K, Cohan A. SCIBERT: A pretrained language model for scientific text. EMNLP-IJCNLP 2019-2019 Conference on Empirical Methods in Natural Language Processing and 9th International Joint Conference on Natural Language Processing, Proceedings of the Conference, 3615–3620, 2020.

[98] Swain MC, Cole JM. ChemDataExtractor: A toolkit for automated extraction of chemical information from the scientific literature. J Chem Inf Model, 2016, 56, 1894–1904.

[99] Mavračić J, Court CJ, Isazawa T, Elliott SR, Cole JM. ChemDataExtractor 2.0: autopopulated ontologies for materials science. J Chem Inf Model, 2021, 61, 4280–4289. issn: 1549-9596.

[100] Tabor DP et al. Accelerating the discovery of materials for clean energy in the era of smart automation. Nat Rev Mater, 2018, 3, 5–20. issn: 20588437.

[101] Huo H et al. Semi-supervised machine-learning classification of materials synthesis procedures. Npj Comput Mater, 2019, 5, 1–7. issn: 20573960.

[102] Kim E et al. Machine-learned and codified synthesis parameters of oxide materials. Sci Data, 2017, 4, 170127. issn: 20524463.

[103] Kim E, Huang K, Kononova O, Ceder G, Olivetti E. Distilling a materials synthesis ontology. Matter, 2019, 1, 8–12. issn: 25902385.

[104] Kuniyoshi F, Makino K, Ozawa J, Miwa M Annotating and extracting synthesis process of all-solid-state batteries from scientific literature. LREC 2020-12th International Conference on Language Resources and Evaluation, Conference Proceedings, 1941–1950, 2020.

[105] Kim E, Huang K, Jegelka S, Olivetti E. Virtual screening of inorganic materials synthesis parameters with deep learning. Npj Comput Mater, 2017, 3. doi: 10.1038/s41524-017-0055-6, issn: 20573960. http://dx.doi.org/10.1038/s41524-017-0055-6.

[106] He T et al. Similarity of precursors in solid-state synthesis as text-mined from scientific literature. Chem Mater, 2020, 32, 7861–7873. issn: 15205002.

[107] Curtarolo S, Morgan D, Persson K, Rodgers J, Ceder G. Predicting crystal structures with data mining of quantum calculations. Phys Rev Lett, 2003, 91, 1–4. issn: 10797114.

[108] Van Der Maaten L, Hinton G. Visualizing data using t-SNE laurens. J Mach Learn Res, 2008, 9, 9, 2579–2605. issn: 15729338.

[109] Noh J et al. Inverse design of solid-state materials via a continuous representation. Matter, 2019, 1, 1370–1384. issn: 25902385.

[110] Pickard CJ, Needs RJ. Ab initiorandom structure searching. J Phys Condens Matter, 2011, 23, 053201.

[111] Oganov AR, Glass CW. Crystal structure prediction using ab initio evolutionary techniques: principles and applications. J Chem Phys, 2006, 124, 244704.

[112] Jumper J et al. Highly accurate protein structure prediction with alphaFold. Nature, 2021, 596, 583–589.

[113] Roter B, Dordevic SV. redicting new superconductors and their critical temperatures using machine learning, Physica C: Superconductivity and its Applications. 2020. 575. https://doi.org/10.1016/j.physc.2020.1353689.

[114] Stanev, V., Oses, C., Kusne, A.G. et al. Machine learning modeling of superconducting critical temperature. npj Comput Mater 4, 29 (2018). https://doi.org/10.1038/s41524-018-0085-8.

Gustavo Guzman

6 Industrial materials informatics

"The materials available to society have defined the very substance of civilization" [1], is the opening phrase by Dr. Taylor Sparks, an associate professor of materials science and engineering at the University of Utah, during his TEDx talk on materials informatics (MI) in 2019. His statement is particularly relevant now, when society faces major challenges with sustainability and the threat of climate change.

Materials informatics, the application of data-driven modeling and informatics to material science, has gained increased attention as an exciting new avenue to rationalize materials discovery and tackle some of the largest challenges in society. This has led to an immense increase in academic publications, especially since the launch of the Materials Genome Initiative in 2011 [2].

Not every new exciting academic trend gets translated into industrial applications with direct societal impact. Sometimes, the translation from academia to industrial applications is thwarted by unexpected factors. For example, we saw lots of excitement and thousands of publications, over at least three decades, on polymer nanocomposites. And yet, nanoscale reinforcing agents at commercially deployable scales have been only sporadically successful to date [3].

So, we can ask ourselves, in the 11 years since the launch of the Materials Genome Initiative, has materials informatics found a relevant translation from the academic laboratory into the industrial sector? The short answer is yes!

What are the societal and commercial pressures compelling the industrial sector to adopt materials informatics? What are the specific materials informatics workflows which have found commercial success? What are the barriers for adoption of the technology in industrial laboratories?

This chapter aims at answering the above questions. The answers are constructed considering not only published literature but also my personal experience leading AI digitalization efforts at large chemical companies and our collective lessons at Citrine Informatics, where we have spearheaded the commercialization of materials informatics for the last decade.

Gustavo Guzman, Citrine Informatics, USA

https://doi.org/10.1515/9783110738087-006

6.1 The societal and commercial case for the industrial adoption of materials informatics

Three factors must come together for the successful implementation of a new technology in an old industry: Technological readiness – it is possible –, commercial drive – it is profitable, – and societal demands – it is the right thing to do. In other words, corporations will have a hard time adopting new technologies if they are either too expensive to implement at scale, do not produce meaningful return on capital, or if doing so will significantly affect the organization's reputation. Below we examine the societal and commercial drivers for the commercial adoption of materials informatics.

6.1.1 The societal case for materials informatics: a warming planet and increased societal awareness

When we hear about climate challenge and the pressing need to reduce our greenhouse gas (GHG) emissions, some sectors tend to get more societal attention and media coverage than others. The transportation and energy sectors, along with the big oil companies, get most of the bad press. Several technological solutions could enable these sectors to dramatically reduce their carbon footprint. Electric cars, battery technology, solar and wind energy, and even a resurgence in nuclear power are all possible solutions [4].

How about materials? Materials make up, quite literally, everything, and are an integral part of both old and future technological solutions for all sectors. From the pipes of the oil refinery to the blades of wind turbines and the photovoltaics cells of solar farms, materials are a key component of all technological developments. You will need lots of steel, glass, and polymers whether your car has an electric motor or a V8. In 2008, the national academy of engineering identified the so-called 14 grand challenges in engineering. These are the challenges we would have to solve to keep progressing as a species. Well over half of them would require the discovery and deployment of new advanced materials [5]. Thus, whichever solutions win the race of decarbonization in the mobility, energy, agricultural, or food tech sectors will require new and better materials.

The key question is then, how do materials themselves fair in the GHG generation picture? In a word, terribly. Manufacturing of materials accounts for 31% of the 51 billion tons of carbon dioxide (CO_2) released globally every year [4]. Furthermore, material-related emissions comprise half of the industrial GHG output, eclipsing mining and general energy supply [6, 7].

As seen in Fig. 6.1, GHG emissions from material production increased by a whopping 120% from 1995 to 2015. This increase is equivalent to 6 billion metric tons

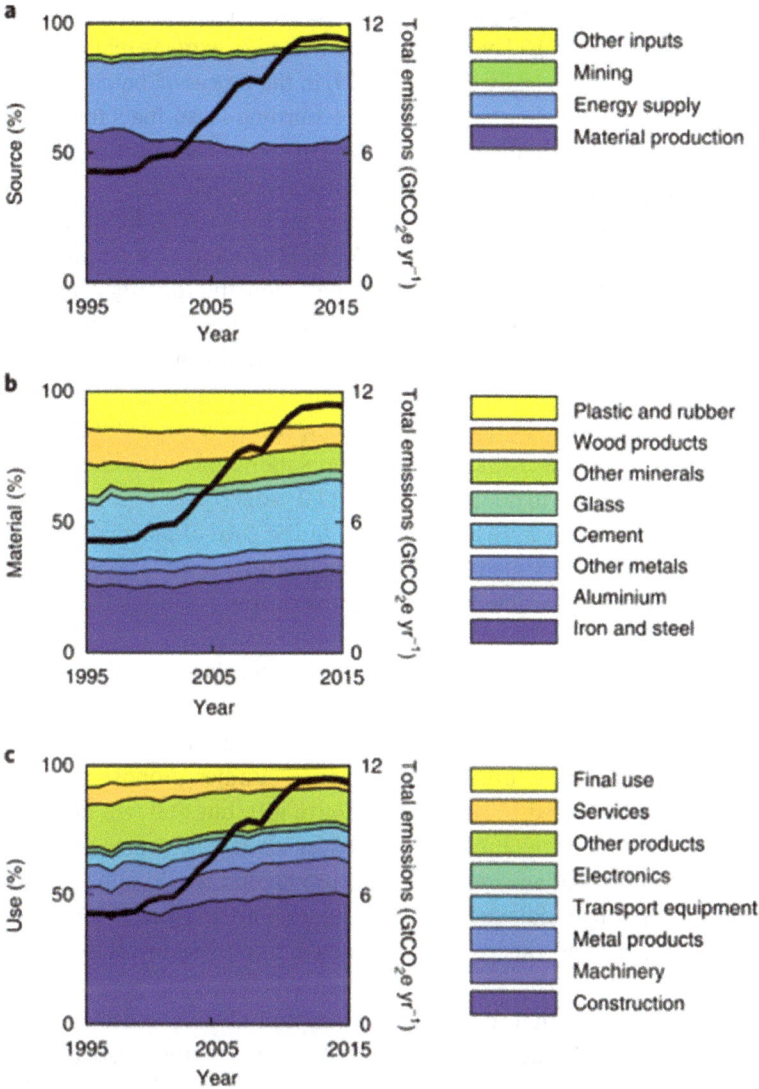

Fig. 6.1: GHG emissions from material production. (a) Emitting process, (b) material class, and (c) carbon footprint of materials by industry. Adapted from [6].

of CO_2-equivalent (GtCO$_2$e)[6]. *CO_2 equivalents* are defined as a metric for GHG emissions where the emissions of methane, nitrous oxide and other minor GHGs are converted to an equivalent amount of CO_2 that would produce a comparable amount of climate change over a 100-year time horizon [6]. Steel (31%) and cement (24%) are the major GHG emitters, with plastics (13%) coming in a distant third. Plastics of course have different but equally concerning environmental issues related to end-of-life disposal, accumulation in the ocean and microplastics.

For cement, the production of CO_2 is inherent to the chemical process involved in its production. Limestone ($CaCO_3$), clay, and sand are heated to extreme temperatures to generate cement clinker, chemically producing CO_2 in the process. Chemically produced CO_2 is added to the CO_2 already generated by burning fossil fuels to provide heating for the process. All in all, about 1 ton of CO_2 per ton of cement is generated. About 4 billion tons of cement are produced globally each year [4].

Steel has the same issue; CO_2 is not only generated by burning fossil fuels to provide heating in the process but also chemically. Making a ton of steel produces 1.8 tons of carbon dioxide. By 2050, the world will be producing roughly 2.8 billion tons of steel a year. That amounts to 5 billion tons of carbon dioxide every year [4].

Construction and urbanization will be the leading growth drivers for cement and steel. Currently, more than 50% of the world's population lives in cities. By 2050, 2.5 billion new residents, most of them in Asia and Africa, will have moved into urban areas, rising the figure to 66% [8].With a growing population and increasing wealth, demand for material is expected to double in the next 40 years [7].

Plastics come in third place in GHG emissions. In the case of plastics, the carbon from refined oil becomes trapped as part of the material itself. The emissions generated in plastic production come mostly from the burning of fossil during the production process.

Critically, although materials production is energy intensive, it is also already largely energy efficient [7]. Production processes for the major material groups have already been subject to many decades of energy optimization efforts motivated by costs. There are remaining opportunities for efficiency (studies propose that efficiency could be further improved by, at most, 25–40% [6]), but they are not sufficient to meet emissions reductions targets proposed by climate scientists [7, 9, 10]. More specifically, such level of efficiency gains in production would allow the expected doubling of global materials demand to come with no increase in emissions [6]. Our target, however, is not to keep GHG emissions at its current levels but rather bring its current number to zero by 2050 [4].

McKinsey & Co. estimates that, optimistically, carbon dioxide emissions from cement production could be reduced by 75% by 2050. However, they also add that only a small portion, around 20%, is likely to come from operational efficiency improvements. The remainder will need to come from technological innovation in materials science [11].

Finding technological solutions to decrease the CO_2 emissions generated to produce materials in the next 40 years will be one of the largest societal and political challenges to ever face materials industry.

The materials industry will be required to solve multiple problems simultaneously:
- Engineer less CO_2-intensive substitute materials with comparable performance and available in the same massive quantities.
- Develop materials and chemicals capable of safely and cheaply capturing massive amounts of CO_2.

– Reduce our total demand for materials by pursuing material efficiency [7]. Material efficiency implies to continue to provide the services delivered by materials, with a reduction in total production of new material. Evidence suggests that, on average, one-third of all material use could be saved if product designs were optimized for material use rather than for cost reduction [7, 10, 12].

The issue is of course that the traditional approach to material discovery is to play it safe. The materials industry has been comfortable looking at already known high performing families of materials and making slight changes to the composition and microstructure to generate new products. Local optimization and incremental discovery have kept the industry afloat for the past several decades.

All these issues will weigh on the chemicals and materials industry in the next several decades, with more regulations and societal pressure accumulating every year. The industry will be forced to invest in technologies to dramatically accelerate the rate of innovation. This is where materials informatics comes in.

6.1.2 The commercial case for materials informatics

Although we would hope this would not be the case, the industrial sector rarely makes large technological investments simply "for the good of humanity." Our decarbonization targets are several decades away and by themselves are not likely to significantly alter the immediate agenda of CEOs concerned with quarterly reports and prices per share.

The good news is that there are, today, major commercial drives for the materials industry to invest in materials informatics.

The reality is that the materials and chemicals industry is at a crossroads. The following are some of the commercial factors contributing to the need for investment in materials informatics.

6.1.2.1 Commoditization

"Innovation to develop new differentiated – and thus specialty – products has become a game of inches. With the exception of innovative crop protection, we would be hard pressed to name a single chemical blockbuster developed in the last ten years" [13], concludes McKinsey & Co. in a 2017 report on the state of the chemical industry and its future.

The reality is that the rate of discovery and commercialization of new materials and chemicals has been slowing down since at least the 1960s [14]. Figure 6.2 illustrated this phenomenon. During the first half of the twentieth century, and especially after the end of World War II, the materials industry saw an explosion of new

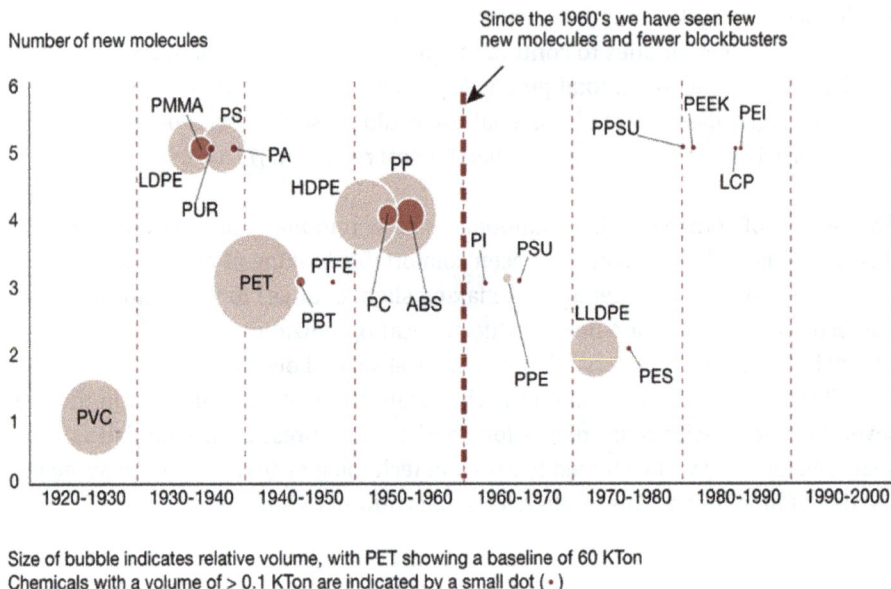

Size of bubble indicates relative volume, with PET showing a baseline of 60 KTon
Chemicals with a volume of > 0.1 KTon are indicated by a small dot (·)

Fig. 6.2: Decline in number of new molecules and blockbusters. Adapted from [14].

blockbuster synthetics materials. The objective during these early years was the substitution of natural materials. In the 1970s, the industry shifted from substitution to the synthesis of custom materials with tailored functionality [8]. During all this time, the industrial research and development model (R&D) remained largely unchanged; researchers synthesize new materials and molecules, marketing finds applications for the new products and markets them to customers [8].

As the well of new molecules and materials began to dry up, innovation strategies shifted toward greater customer feedback. Today customers often dictate specifications for materials, which are then developed and produced.

In practical terms, chemicals and materials companies are now facing a progressively tougher business environment as larger segments of the industry become commoditized. Specialty chemical players are struggling to create innovative products that offer differentiated value [15]. Commoditization leads to margin erosion, as prices become cost-driven instead of value-based, and volatility increases. Figure 6.3 illustrates commoditization across several chemical sectors [16].

The number of new chemical producers has increased as well (mostly in emerging markets), and thanks to the fact that chemical production technology has become more broadly available, new producers have been able to build capacity additions ahead of market growth, depressing prices and eroding margins [15]. As new competitors from countries with lower production costs entered the market, being able to offer similar products to those of the incumbents but at lower prices, the situation for long-term established players has become even more difficult.

The commodity frontier continues to move to the right.

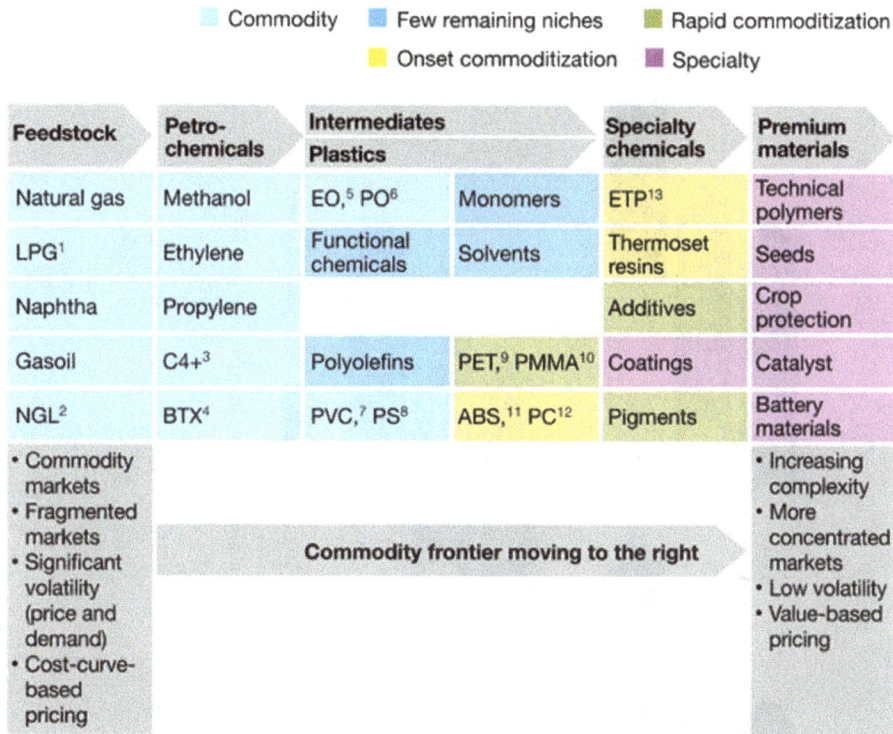

| Commodity | Few remaining niches | Rapid commoditization |
| Onset commoditization | Specialty |

Feedstock	Petro-chemicals	Intermediates Plastics		Specialty chemicals	Premium materials
Natural gas	Methanol	EO,[5] PO[6]	Monomers	ETP[13]	Technical polymers
LPG[1]	Ethylene	Functional chemicals	Solvents	Thermoset resins	Seeds
Naphtha	Propylene			Additives	Crop protection
Gasoil	C4+[3]	Polyolefins	PET,[9] PMMA[10]	Coatings	Catalyst
NGL[2]	BTX[4]	PVC,[7] PS[8]	ABS,[11] PC[12]	Pigments	Battery materials

- Commodity markets
- Fragmented markets
- Significant volatility (price and demand)
- Cost-curve-based pricing

Commodity frontier moving to the right

- Increasing complexity
- More concentrated markets
- Low volatility
- Value-based pricing

Fig. 6.3: Commoditization in chemicals. Adapted from [16].

In the west, where markets are mature, this margin erosion has been significant enough as to erase most of value pool generated during the 2000s decade, even with advantaged feedstocks in places like the USA and a 4% compounded annual growth rate (CAGR) [15, 16].

6.1.2.2 The deteriorating feasibility of traditional materials R&D

In the previous section, we described how the slowing rate of new blockbuster materials resulted in commoditization and margin erosion. To make matters worse, not only are new materials being discovered less frequently, but the time from discovery in the laboratory to deployment in the marketplace has remained vexingly long. Indeed, as we can see in Fig. 6.4, it can take in between 10 and 20 years *after* the already costly and risky process of invention for a new material to be fully scaled and commercialized [8, 17].

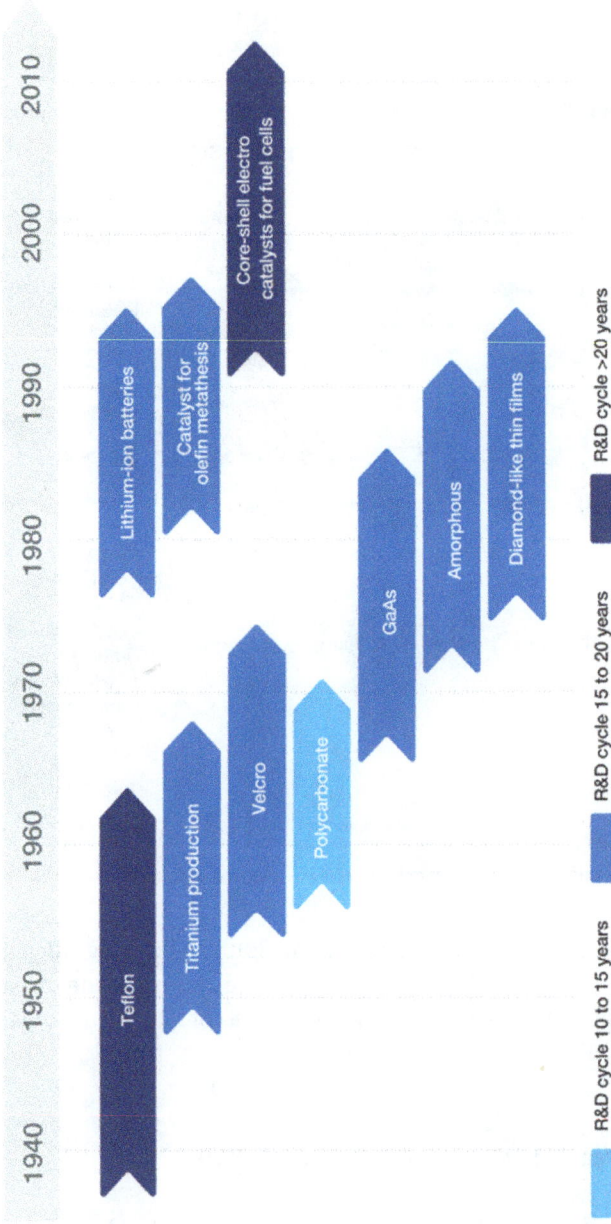

Fig. 6.4: R&D lifecycles for selected twentieth century materials innovations. Adapted from [8].

While materials development lags, consumer product development cycles have only become faster. In a 2001 report, the National Research Council highlighted that taking a new consumer product from invention to widespread adoption typically takes 2 to 5 years [17]. 2001 was over two decades ago, and development cycles have only accelerated since then. With faster product development comes more stringent end-user requirements and heightened expectations.

It has become painfully clear that new materials development is a bottleneck for new product development. Investors and executives have become increasingly impatient with slow rates of innovation, plateauing margins, and the low effectiveness of the current approach to materials R&D. It should not come as a surprise then, that downstream companies, in an attempt to debottleneck their product lines, have begun to create internal materials development capabilities that are directly competing with global chemical companies [8, 17].

As the demands for quick and custom solutions keep mounting, materials R&D will be required to move from the traditional linear model to a more complex platform model in which end-user needs are paramount [9]. While today customers choose materials from a catalog of material properties, in the future, we can expect a shift toward specialization and a broader portfolio of new materials tailored to specific applications. For companies in the materials space, success will hinge on the ability to quickly identify, design and develop the new materials that can best meet the shifting priorities of their customers [18].

6.1.2.3 Increased regulation and evolving consumer preferences

Earlier, we highlighted the impact materials and chemicals have on our warming climate and our social responsibility to find technological solutions. Customers are now demanding more sustainably sourced – and manufactured – products and asking companies to limit the use of fossil fuels and non-renewable material sources. The world economic forum highlights that 55% of global online consumers across 60 countries say they are willing to pay more for products provided by companies that are committed to positive social and environmental impact [8].

From a more pragmatic lens, consumer environmental concerns are usually translated into regulations. As new and more stringent regulations emerge, materials companies will be forced to adapt and innovate if they do not want to be regulated out of business [19].

As a final note for this section, the COVID-19 pandemic has exposed the vulnerabilities of complex global chemical supply chains built on lean manufacturing principles. The need to design smarter, stronger, and more diverse supply chains has been one of the main lessons of this crisis. Supply chain planning is likely to transform from minimizing fixed and variable long-term supply-chain costs to striving for cost-efficient robustness, redundancy, and diversification. For materials and

Customer Requirements	Consumer Expectations	Sustainability Demands	Evolving Regulations	Uncertain Trade Landscape

Fig. 6.5: The commercial case for materials informatics.

chemical companies, an evolving supply-chain landscape adds to the pressure for agility in material design and formulation.

Let's summarize why the materials industry is in desperate need of materials informatics (see also Fig. 6.5):

- The materials and chemicals industry is a major contributor to climate change. Radical new materials innovation, going far beyond manufacturing efficiency improvements, will be required in the next 40 years to dramatically reduce GHG emissions as demand for new materials doubles.
- In the past decades, the materials and chemicals industry has seen fewer and fewer new blockbuster molecules, and innovation as become a game of inches. This trend has led to wave of commoditization, eroding margins and investor confidence.
- Even after discovery, it can take decades to bring a new material to market. As consumer products' development cycles accelerate, materials development has become a bottleneck for new product development. Materials companies are evolving to become more product centric – instead of molecule centric – and are required to quickly respond to evolving customer requirements.
- Sustainability is shifting consumer expectations and regulatory frameworks. Companies will be forced to quickly adapt to new environments.

6.2 Industrial materials informatics

The common trend for all the points highlighted in the above summary is one: Speed! The materials and chemicals industry is in desperate need of an agility boost. It needs to be able to react to increased customer requirements, evolving regulations, commoditization, and the threat of climate change by quickly developing new targeted solutions.

In the following, we will describe how chemical and materials companies are applying materials informatics to accelerate their materials development, today. Although I cite published literature throughout, most of the insights presented come from the lessons learned at Citrine Informatics while collaborating with leading chemicals and materials companies in many industrial use cases.

6.2.1 Sequential learning for accelerated materials discovery

Our experience shows that machine learning (ML) models in industrial settings are more useful as guides for an iterative sequence of experiments, as opposed to one-time screening tools to evaluate an entire search space to find high-performing materials. As we will see, the objective of industrial materials informatics is the reduction of failed development cycles[20]. Examples of product development cycles for new product development and applications engineering are displayed in Fig. 6.6. Companies are looking to identify a new material that meets customer specifications, develop a new application/formulation, or improve an existing material using as few physical experiments as possible (see Fig. 6.7). We therefore define the ideal application of machine learning in materials discovery as experiment prioritization, rather than materials property prediction [21].

We have coined the term sequential learning (SL) to refer to the process used to accelerate materials discovery projects by reducing the number of experiments required to find a material that meets a given set of target specifications. Mathematically, the process focuses on choosing the parameter settings for an experiment – or series of experiments – to either maximize information gain or move toward some optimal parameter space. A machine learning model – random forests, Gaussian processes, and neural networks are the most popular – is trained on a small initial dataset. The trained model is subsequently paired with an optimization algorithm and an acquisition function to search over a search space of possible experimental parameters for the most promising material candidates and suggest them for experimentation. A material scientist then performs experiments on the candidates suggested by the model, and later retrains the model after adding the new experimental results to the training set [22, 23].

An acquisition function is a decision rule used to rank potential material candidates. Such criterion allows the material scientist to explore the "exploration-exploitation" trade-off and create a selection strategy. This way, the algorithm can seek candidates with improved properties but also try "risky" candidates to improve its model and enable later discoveries. [20] See Fig. 6.8 for a flow diagram of the process.

Formally, a real response function, – the property of interest being measured in the experiment, which is expensive to evaluate – $f(x_{n_0})$ is calculated by a surrogate ML model $\hat{f}_0(x_{n_0}, \theta)$, where x_{n0} are the initial experimental material parameters – material descriptors – and θ are the surrogate ML model parameters. The initial

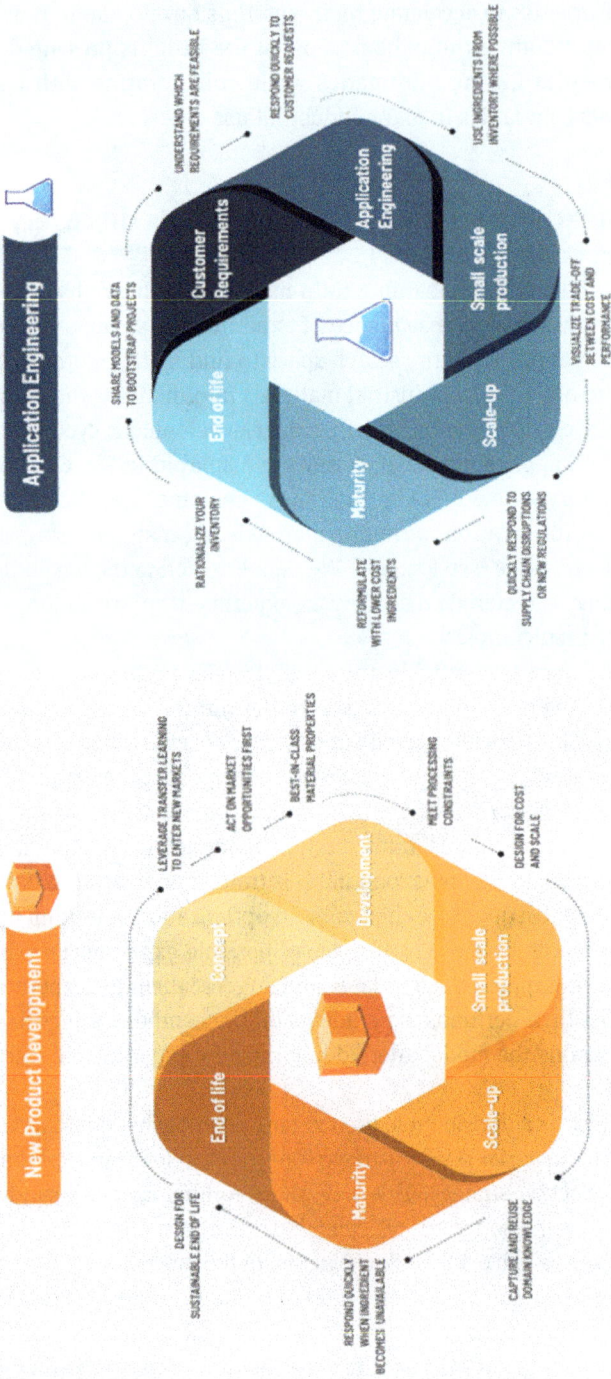

Fig. 6.6: Material development cycles.

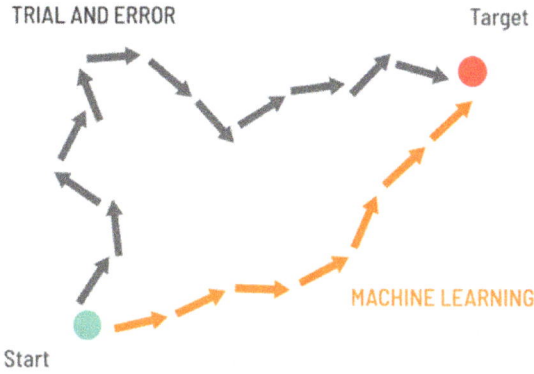

TRIAL AND ERROR

Target

MACHINE LEARNING

Start

DATA-DRIVEN RESEARCH STREAMLINES
THE NUMBER OF EXPERIMENTS NEEDED TO
OPTIMISE A MATERIAL.

Fig. 6.7: Reduction in the number of required experiments through machine learning.

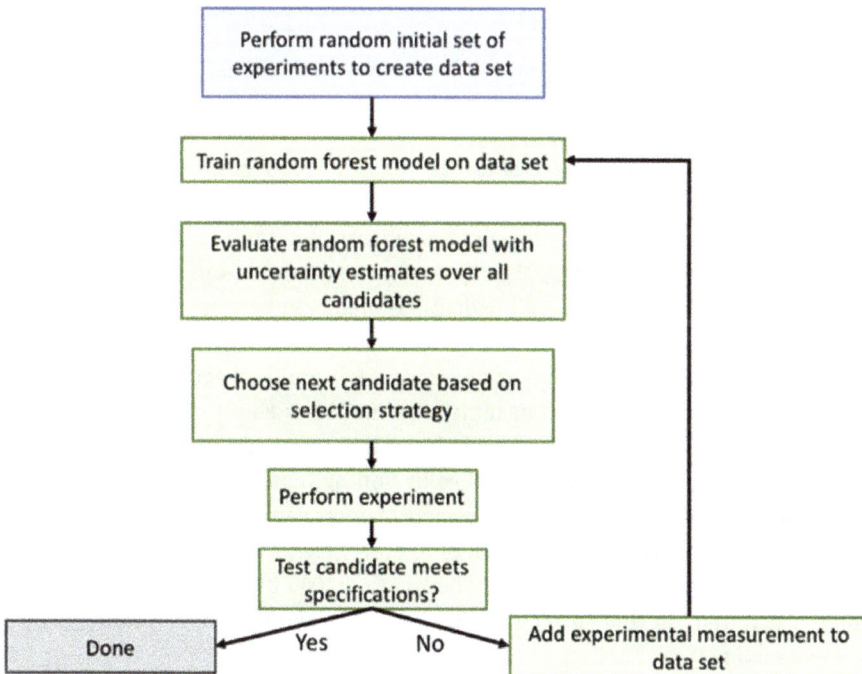

Perform random initial set of experiments to create data set

Train random forest model on data set

Evaluate random forest model with uncertainty estimates over all candidates

Choose next candidate based on selection strategy

Perform experiment

Test candidate meets specifications?

Done

Yes

No

Add experimental measurement to data set

Fig. 6.8: The sequential learning process. Adapted from [22].

surrogate function $\hat{f}_0(x_{n_0}, \theta)$ is learned from a training set and, along with an acquisition function, is used to generate new candidates $x_{n_0+1}, \ldots, x_{n_0+n_1}$. The new candidates are characterized and added to the dataset to train a new surrogate model $\hat{f}_1$. While a common objective in machine learning would be to improve the model $\hat{f}$, our objective is to identify an improved material x with as few physical experiments/SL cycles as possible [20, 22].

After having run many industrial materials informatics projects, we have identified a set of particularities and challenges for applying the SL framework to commercial materials selection, improvement, and discovery. These particularities are related to both the unique conditions of the materials and chemicals space but also to the objective of finding new candidate materials instead of doing a one-shot prediction. Table 6.1 summarizes the particularities of materials informatics in industrial settings and compares them with more general applications of machine learning.

The first particularity we have found is that the objective of identifying new promising material candidates with optimized properties is distinct from the traditional aim of machine learning of improving global model performance [20, 21]. In fact, minimizing the global model error can be at odds with acquiring optimal candidates and realizing a model with the lowest global error does not guarantee optimal candidate discovery.

Finding the optimal material for a given application through the SL process requires a metric which the acquisition function can use to rank material candidates and propose the next round of experiments. This is not an issue in the case when a single property is being optimized since the algorithm can simply perform a scalar ranking. However, this is rarely the case in industrial settings. Industrial scientists are commonly faced with the challenge of having to simultaneously optimize multiple objective properties to comply with customer, process, and product requirements. In as many cases, there are inevitable tradeoffs in between desired property targets. Typical examples include the stiffness vs. impact resistance inverse relationship and the strength/ductility tradeoff. In such multi-objective cases, there is no simple way of ranking different material candidates as no single candidate is objectively better in every dimension and no single optimum point is attainable. Consequently, the principal challenge lies in finding materials at the *Pareto front*, which is the surface over which any incremental improvement in one material property requires a commensurate reduction of another (e.g., a 15% reduction in impact strength of a thermoplastic olefin may be worth a 10% increase in stiffness) [17]. Formally, materials in this surface are said to be Pareto-optimal solutions [24] or *non-dominated* points.

Since our thesis is that the value of materials informatics for the chemicals and materials industry is in the reduction in failed development cycles, it is critical to

Tab. 6.1: Particularities for the application of machine learning in materials informatics.

	Traditional MI Applications	Materials MI Applications
DATA TYPE	Often standardized	Rarely standardized
DATA VOLUME	Big, dense (up to ~10^8 examples)	Small, sparse (~10^2 examples)
ESTABLISHED DOMAIN KNOWLEDGE	Not applicable—rely on data to learn patterns	Must be physics-aware
DATA REPRESENTATION	Can often be optimized by algorithms	Requires deep domain knowledge
PREDICTION TASK	Accurately pattern-match common cases	Predict unusual or "extreme" materials
SAMPLE BIAS	Often present	Experiments correlated; negatives stigmatized
UNCERTAINTY IN DATA AND MODELS	Usually unimportant	Always important
INTERPRETABILITY	Usually unimportant	Often required by scientists & engineers

define efficient search strategies to find Pareto-optimal materials, as to minimize the number of physical experiments. A wide variety of metrics [20] for acquisition functions in multi-objective optimization have been proposed, which we will not cover in detail here. Instead, we will focus on how the characteristics of the problem at hand influence the decisions of the material scientist in an industrial setting. As mentioned above, one of these characteristics is that the global error becomes an unreliable metric to evaluate the machine learning model's ability to find Pareto-optimal candidates.

Figure 6.9, as reported by Ling et al, presents a relevant example for thermoelectric materials. In the top graph, we see the evolution of a global error measure as a function of the number of SL iterations, while the bottom plots the number of optimal candidates found. The measure used is the mean non-dimensional model error (MNDE) [20]. The global error is presented for two different multi-objective acquisition functions. Both acquisition functions aim at acquiring non-dominated material candidates. We can see that the maximum probability of joint exceedance (MPJE) function has a better performance error-wise but also is able to find fewer optimal candidates, while the maximum probability non-dominated (MPND) function presents the opposite case.

Ling introduces shell error [20] as a new metric that scopes the error evaluation to particularly bands about the Pareto frontier (Pareto shells) and provides a better measure of improved candidate discovery. Since the global error is found to not be predictive of candidate discoveries, industrial scientist looking to improve multiple properties should assess both the global and shell error when evaluating the model's ability to produce and effectively rank material candidates. Interestingly, we have found that having an acquisition function that accurately maps the Pareto frontier is less critical for early stages of the SL process. Accordingly, if a given project has a large set of uncharacterized material candidates, the choice of multi-objective acquisition function is less critical. If, however, the design space has been more thoroughly explored then accurately capturing the Pareto frontier in the acquisition function's logic becomes important [20].

The above conclusions have important consequences for different materials classes and chemical sectors. While in certain material sectors search spaces might remain highly unexplored – new materials, specialty chemicals, formulations – in more commoditized sectors industrial scientists might have been pursing incremental improvements for decades. This means that scientists working on long-run discovery of non-dominated candidates should place particular attention on capturing Pareto frontier with greater fidelity [20].

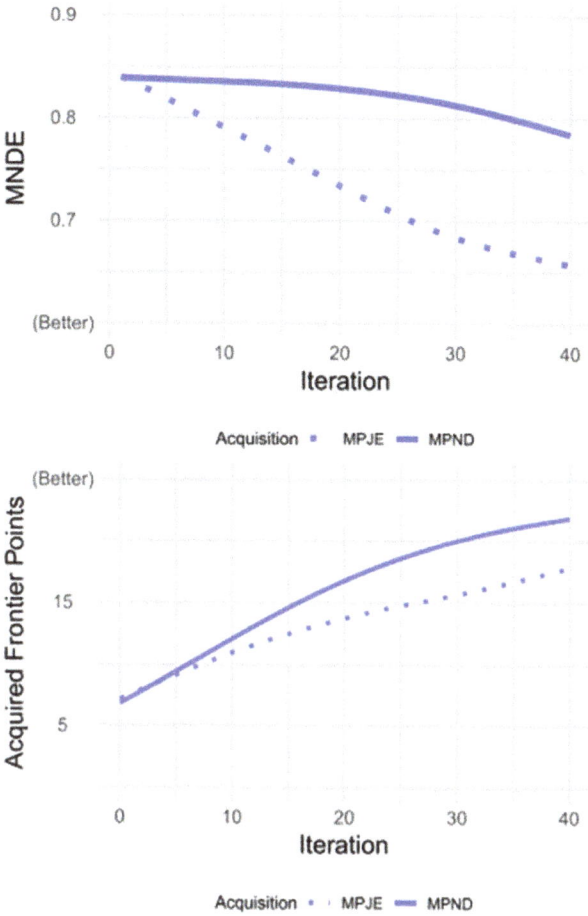

Fig. 6.9: Best global error does not guarantee optimal candidate discovery. Shown are global prediction errors (MNDE, top) and number of optimal candidates found (bottom). Adapted from [20].

6.2.2 Well-calibrated uncertainty estimates

The second particularity we have identified in the application of materials informatics and sequential learning to industrial settings is the need to for effective uncertainty quantification (UQ) and well-calibrated uncertainty bounds on model predictions.

At the outset, the assessments above regarding the identification of Pareto-optimal materials through SL rely on having well-calibrated uncertainty bounds on model predictions [21]. For example, if a model predicting impact strength has no uncertainty bounds, it is not possible for the scientist to know if the model is evaluating a known area of the input space or extrapolating wildly, and therefore cannot balance exploration and exploitation. Uncertainty estimation becomes even more relevant in industrial settings where sparse and small datasets are the norm, which usually results in initial models with high uncertainty [22, 25]. In these cases, looking for high uncertainty, "risky," candidates that provide high informational gain can be as useful as focusing on finding optimal candidates in known spaces.

In the industrial application of materials informatics, effective uncertainty quantification is not only relevant for the successful application of sequential learning but also, vitally, for the enablement of executive data-driven decisions. A significant way chemicals and materials companies are using uncertainty quantification in data-driven decisions for materials development is through search space evaluation and visualization.

So far, we have highlighted how materials informatics can aid industrial scientists in minimizing the number of failed development cycles by choosing optimal experiments from a search space. We have not mentioned however, how to choose the proper search space to begin with. The quality of the search space might make discovering an improved material trivial – if all the untested materials are non-dominated– or close of impossible if the opposite is true.

6.2.3 Search space evaluation

In real industrial problems, one of the main challenges is the immensely large space of potential experiments that a R&D organization could pursue in any given project. If solving a single SL project is like finding a needle in a haystack, then the real challenge chemicals and materials companies face is that of finding if there any needles in an entire field of haystacks, not knowing if any given haystack has a needle in it [23]. Each metaphorical haystack is as a search space, a group of possible experiments where one might find optimal materials. It would be extremely useful for the industrial scientist then, to know from the start if for a given design space is likely to take 10, 100, or 1,000 experiments to find an optimal material [23], or even better, if there are no improved materials to find at all!

Determining the quality of a search space has been shown to be a great tool for management, as it provides an effective way of promoting or killing projects. Bad projects can die sooner and good projects and be further investigated.

Ling et al proposed the fraction of improved candidates (FIC) as a metric for the evaluation of the quality of a given design space. The FIC is defined as the fraction of candidate materials in the design space that perform better than the best candidate material present in the training data (i.e., the fraction of the haystack that are needles) [23]. Fundamentally, we have found that design spaces with a low FIC will require, on average, many iterations to find an improved candidate, while design spaces with high FIC will require a comparatively lower number of trials to find an improved candidate. Interestingly, and in agreement with previous results, it was found that model quality, unlike search space quality, is not strongly tied to sequential learning success [20, 23].

Ling highlights that in the real scenario of searching for an undiscovered material, the exact value of FIC is unknown a priori. However, she proposes that we can find predictive metrics correlated with the FIC in order judge design space quality.

The first metric found to be highly correlated to the true FIC was the PFIC score, defined as the fraction of design space candidates that are predicted by the model to have improved performance over the best training data point

$$PFIC = \frac{N_{\rho(x_i) > b}}{|X|} \tag{6.1}$$

where X is the search space, x_i represents each candidate in the search space, $\rho(x_i)$ is the predicted performance of that candidate and b is the performance of the best training point. $|X|$ is the design space size, and N_α is the number of candidates satisfying the condition α. The numerator in equation 6.1 represents the number of search space candidates predicted to be improvements over the best training data point while the denominator represents the search space size. Equation 6.1 is set for a property maximization, in the minimization case (1-PFIC) should be calculated [23].

The PFIC score can be used to identify high quality search spaces, as it identifies SL projects that are likely to succeed, "The haystack is full of needles!"

The second metric reported to be correlated with the true FIC was the cumulative maximum likelihood of improvement (CMLI). The CMLI is defined as the predicted probability that at least one candidate, out of top n candidates with the highest likelihoods of improvement, performs better than the best training data point, where n is a tunable parameter [23]. This metric relies on the machine learning algorithm generating predictions with well-calibrated uncertainty bounds, as the calculation of the likelihood of improvement L for a given candidate depends on the standard deviation estimated by uncertainty. Once L has been calculated for all candidates in the search space and the top n candidates have been identified, the CMLI can be computed by:

$$CMLI = 1 - \prod_{i=1}^{n} (1 - L(x_i)) \tag{6.2}$$

In equation 6.2 we take the product of the likelihoods that the top n candidates x_i are *not* an improvement, $(1 - L(x_i))$.

The CMLI score is evaluating the likelihood that *at least one* of the top n candidates is an improvement over the best candidate in the training set. Ling identified that, contrary to the PFIC, the CMLI score can be used to flag *low-quality* search spaces at the outset of a project, "few needles in the haystack." Critically, both metrics, PFIC and CMLI, can be computed without any extra data acquisition, meaning no sequential learning iterations are required to calculate the predictive search space metrics. Note that the above metrics are limited to the single objective optimization case.

R&D managers at chemical and materials companies are now rationalizing their investment portfolio by evaluating potential projects based on the quality of their search spaces and choosing those which are most *likely* to yield improved materials with the least number of development cycles.

6.2.4 Search space visualization

Search space evaluation is helping managers and executives identify projects with a higher chance of finding improved materials. *Visualizing* the search space, on the other hand, brings a different set of advantages.

Search space visualizations illustrate material properties predicted by a trained machine learning model along with their well-calibrated estimated uncertainties. The visualizations enable the assessment of the likelihood that a search space contains materials with properties located at any given point in a two-dimensional material property "output" space [25]. Figure 6.10, as reported by Ling et al., presents an example of one of such visualizations. In the proposed example, a company is aiming at developing a new organic solvent with high relative polarity and high boiling point. The company has been recently been hit by new regulations which require the solvent to be biodegradable. The engineer in charge of the project is interested in determining to what extend the requirement for biodegradability will affect the possible performance on the properties of interest [25].

Ling described two strategies for visualization of the probability of performance. The first strategy is the maximum joint probability density (MJPD), which describes the probability of achieving a given region in the material property output space given the best performing candidate in the design space. The second strategy is the summed probability density (SPD) which describes the predicted density of candidates in the material property output space [25]. In Figure 6.10, the relevant material properties are displayed in contour plots along the x-y axes while either MJPD (top) or SPD (bottom) are plotted along the z axis using a color map. The contours in the MJPD plot indicate that the predicted achievable performance is similar for both the biodegradable and non-biodegradable versions of the design space. The SPD plot shows that the biodegradable design space has a lower prediction density at higher boiling points when compared to the non-biodegradable, but also a higher prediction density at higher relative polarities. The engineer in charge of the project might then conclude that even with the biodegradability requirements the solvent is likely to achieve the desired performance [25].

Search space visualizations are being used at chemicals and materials companies to understand performance trade-offs under different scenarios and make data-driven decisions. Such decisions comprise whether to include a new ingredient, invest in equipment to broaden the processing window, adapt to evolving regulations, or react to supply chain disruptions.

The evaluation and visualization of search spaces expand the envelop of applicability of materials informatics beyond experiment suggestion and help permeate data-driven methods across the entire research process.

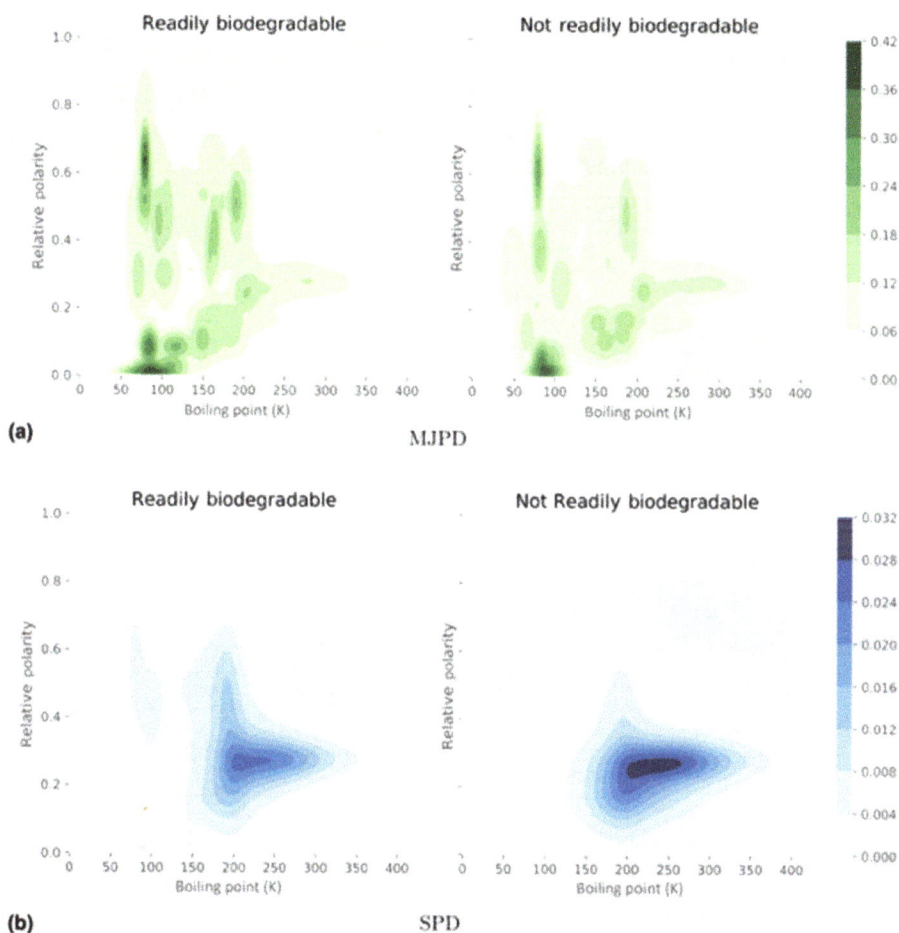

Fig. 6.10: Search space visualization plots for the readily biodegradable and non-readily biodegradable subsets of the search space. (a) is colored by the MJPD metric and (b) is colored by the SPD metric. Adapted from [25].

Summarizing, materials informatics is being employed at leading chemicals and materials companies due to its promise of alleviating the pressures of commoditization, evolving customer demands and sustainability. It does so by:

- Drastically accelerating the material development process by reducing the number of experiments required to find Pareto-optimal solutions using sequential learning with well-calibrated uncertainty estimates.
- Predicting the likelihood of success in materials discovery and rationalizing investment decisions throughout the entire research process using search space evaluation and visualization.

Figure 6.11 illustrates how at-scale implementations of materials informatics employing the workflows described – SL, design space evaluation, and visualization – are allowing chemicals and materials companies to expand their design options, quickly assess a large number of new formulations, optimize them for cost and deploy them to market in record time [26].

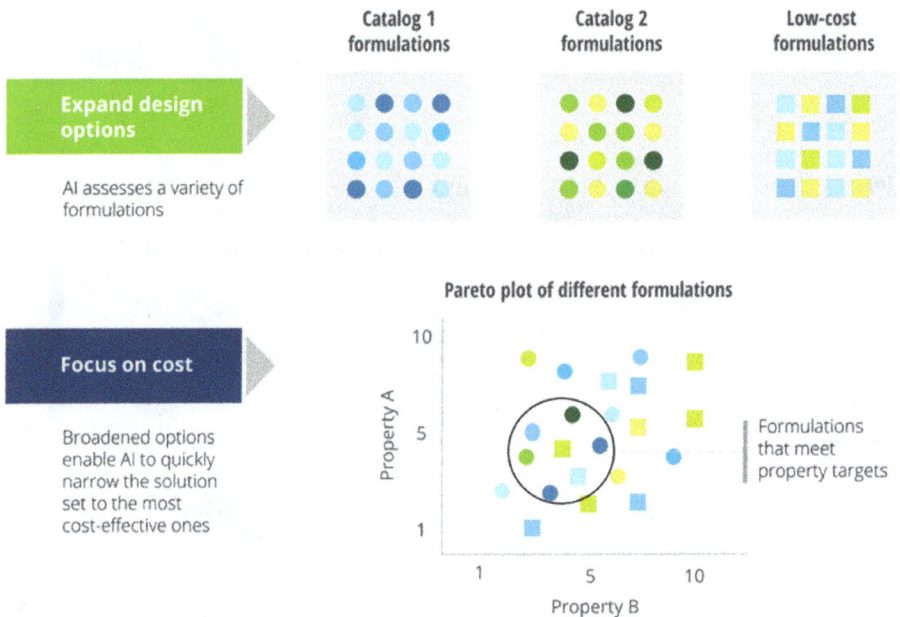

Fig. 6.11: How AI enables rapid reengineering of products for reduced cost and changing supply chains. Adapted from [26].

6.3 The industrial laboratory data-management ecosystem: challenges and value drivers

In the previous two sections, we described why the chemicals and materials industry is primed to adopt materials informatics and the ways in which MI is being used commercially for sequential learning, replace space evaluation and visualization. In this section, we turn to data and the industrial data landscape.

A trend in academic publications in materials informatics is to focus on improving modeling, optimization, and candidate acquisition techniques on large public datasets. When discussing issues with materials data, academia tends to focus, naturally, on academic literature data issues.

For the industrial application of materials informatics however, the quality and quantity of historical data accessible to the firm, the collection, storage and retrieval systems, the data and laboratory management protocols and the accompanying digital ecosystem (production, procurement, logistics) are as important, or more so, than using the right optimization strategy.

The following section describes the value drivers chemicals and materials companies have had, until now, to store and curate their materials data. We then highlight how those value drivers have shaped the current industrial data-management ecosystem and how these legacy systems affect the commercial application of materials informatics.

6.3.1 Historical business drivers for scientific data management: it's all about defense

To begin, let's recall the traditional R&D model; researchers synthesize new materials and molecules, marketing finds applications for the new product and markets them to customers. For the longest time, upper management trusted their scientist to use their intellect and instincts to come up with clever new molecules, as such, scientific data belonged to the individual scientist.

As the industry grew exponentially in the second half of the twentieth century and competition increased, we find the first the business driver for scientific data management, lawsuits. Companies were being challenged by competitors on intellectual property (IP) grounds, pushing them to have more strict measures for scientific record storage. The emergent systems of record were naturally optimized for "low-throughput" human consumption [27] and focused on storing documents annotated with metadata useful for the IP lawyer. If the firm could identify who did what and when, the scientific details need not to be standardized. Record retrieval events were expected to be uncommon and the firm could always bring the scientist in charge of the project into the courtroom.

In the later decades, environmental regulations for the chemicals and materials industry became much more relevant and demanding [28]. Companies were now compelled to use specific protocols for storing their scientific records as to remain in compliance with regulations during audits. Once again, the requirements for the data-management system were to provide structured access to specific records in the unlikely event of an audit. The scientific details need not to be standardized. Reusing and repurposing others' data was not only unfeasible [27] but it also did not have, at least then, a value driver for the corporation.

Finally, as the commoditization frontier crept on, the industry consolidated and margins eroded [16], more companies were forced to search for cost-reduction measures. Laboratory optimization and productivity became more relevant as a result, and data-management systems which enabled laboratories to run more efficiently – to run more samples and make better use of testing equipment and personnel – emerged as a response [29, 30].

Historically then, corporate incentives for investing in scientific data management have been mostly about enabling the defense of the corporation against negative value events. Not unlike buying insurance, companies invested capital in scientific data management systems to lower the expected value of future negative cash flows due to IP litigation or regulation infringement. Incentives for standardization, large-scale analysis, and reuse of the scientific results themselves were non-existent.

Now, let's contrast the historical value drivers with the emergent needs of the industry and the value enabled by materials informatics. MI enables companies to use their own data as a differentiated strategic asset. By using sequential learning, companies can reach material objectives in strategic areas faster than ever before, beating their competitors to market. By rationalizing their R&D portfolio, they can discard or invest further in projects more effectively, becoming much more agile in the process. Finally, the search spaces, models, and historical SL runs become reusable digital assets which allow these companies to increase their return in subsequent projects at marginal cost. Materials informatics is shifting the value drivers for data management from defense to offense, from reducing the possibility of negative cash flow events to enabling direct revenue growth. Figure 6.12 highlights this contrast.

6.3.2 Current landscape of scientific data management

The historical drivers for scientific data management – IP protection, regulatory compliance and cost cutting – have translated into the current industrial data-management landscape.

Figure 6.13 presents the different database and file systems commonly found across the industry. At the bottom, we have individual instrument files and reports that scientists handle on their own on a daily basis as part of their usual workflow. On the second level, we have cloud or on-premise file sharing systems which have

Fig. 6.12: Value drivers for scientific data management: historical vs materials informatics.

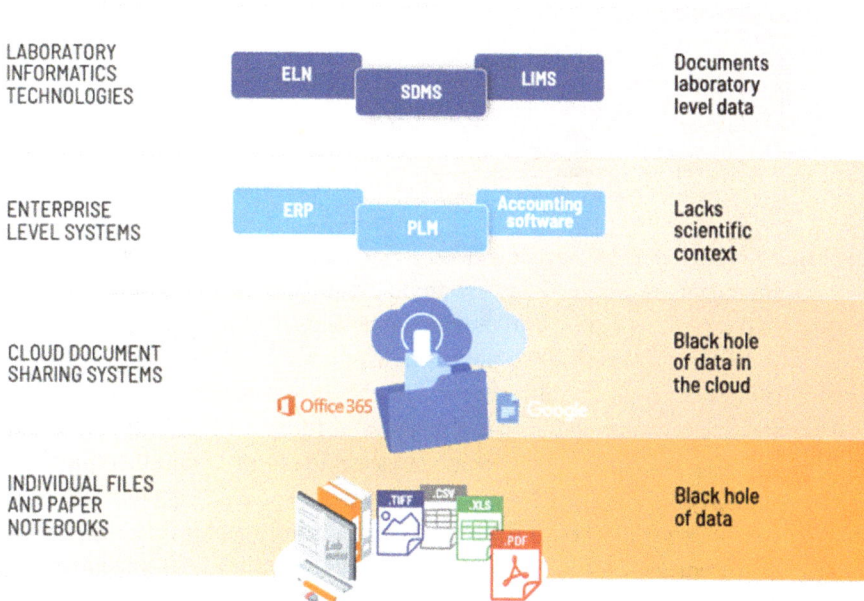

Fig. 6.13: Industrial scientific data management landscape.

become widespread in the last decade, such a SharePoint or Google Drive. Next, we have non-scientific systems material scientists commonly interact with such as enterprise resource planning systems, or production and accounting databases. Lastly, we have laboratory informatics technologies which come in a variety of shapes with overlapping functionality such as electronic laboratory notebooks, laboratory information management systems, and scientific data management systems. Table 6.2 presents the basic differences among these.

Tab. 6.2: Laboratory informatics technologies.

	Overview	Benefits
ELN	Software designed to replace paper laboratory notebooks. Used by scientists, engineers, and technicians to document and share information on research, experiments, and procedures. ELNS are primarily used for data aggregation and IP protection.	• Document experimental design and results • Handle unstructured data and information • Manage research collaboration • Input data as it's generated
SDMS	Software that acts as a document management system (DMS) for capturing, cataloging, and archiving data from different sources, including scientific instruments, analytical applications, or ELN/LIMS. The data in an SDMS is indexed and searchable and can be integrated with other applications.	• Consolidate data and information • Seamless integration with other similar software • Knowledge management
LIMS	The features and uses of a LIMS have evolved over the years. While LIMS started as a software to manage sample workflow and data tracking, many now enable data mining, analysis, and resource planning. They can also integrate with ELNs and other third-party systems.	• Manage structured data and information • Sample tracking and management • Resource and asset management

The first common theme we have identified is that most of the systems dealing directly with scientific data are designed for "low-throughput" human consumption of a single experimental record, as opposed to programmatic access through an application-programming interface (API). Furthermore, single experimental records rarely (never) share a common schema for different experiments or procedures and can vary from laboratory forms to instrumental files, images, and PDFs. The existing tools provide enough metadata to find records associated to specific people or dates, but rarely provide information on the details of the materials science i.e. they provide structured access to unstructured scientific data. A common sight is for companies to have a system that enables scientists to find their own experiments but that at the same time stores idiosyncratic data formats under the hood, with few to none widely adopted data standards [27, 31]. The current status quo in materials data is, of course, fundamentally incompatible with state-of-the-art machine learning and computational workflows that rely on large amounts of structured data. Any attempt at performing data analytics utilizing data from these systems requires for the data to be first scraped or extracted by other software methods [27]. Note that the above statements exclude non-scientific systems such as ERPs, which have been amenable to programmatic queries from the start. LIMS systems can also sometimes effectively export data programmatically, but it is usually within the context of the quality control laboratory where the same measurements are repeated over and over.

Value wise, the reason for having systems that provide structured human consumption of unstructured scientific data is clear, *they are good enough IP protection and regulatory compliance!* Also, they work well enough for the traditional R&D process where small teams of industrial scientist worked in relative isolation to discover with new molecules and throw them over the wall.

There is an important technical factor in here as well. Data workflows in material science vary considerably depending on specifics such as the research focus, data-acquisition techniques, and individual researcher's personal preferences. Until some years ago, the dominant technology for data storage was that of relational databases with rigid schemas. A relational structure might be feasible for a narrowly defined material science case (such as the quality control laboratory), but it quickly becomes brittle as small changes to the research scope can lead to major changes in the structure of the database or break the schema altogether, which leads to long and expensive database refactors or for the construction of a new database to be required for each new study [17].

Presently, multiple solutions exist under the "NoSQL" database tittle, in which data are stored in machine-parsable documents with no externally imposed schema [17]. As we will see, NoSQL database technologies are being used to enable materials informatics at scale. However, it is only as chemicals and materials companies realize the value creation opportunities through data-driven decisions and speeded up materials development, that these new database technologies are adopted. In other words, the way chemicals and materials companies store their scientific data is dictated by the business processes those data enable, and not the other way around. It is only through the adoption of new materials informatics workflows such as sequential learning, search space evaluation and visualization – and the positive cash flows they enable – that data systems can evolve in a commercial context.

Another consequence of the current data-management landscape is that even within a single business unit of a given company, materials data tends to be scattered across multiple silos. The lack of the relevant scientific metadata necessary for the effective replication of a given experiment means that material scientists within the same organization have a hard time verifying the data quality of their coworkers and, as a consequence, tend to trust only those experimental results generated within their own small silo [31, 32]. This data decentralization is more pronounced in companies that have grown through mergers and acquisitions, and it contributes further to the proliferation of data formats and storage methods. This phenomenon is not only conducive to the generation of redundant experimental results but also to the loss of institutional knowledge as senior scientists' insights and intuition are not captured along with their experimental results.

As more companies adopt and scale materials informatics efforts, industrial scientific data systems will adapt to enable the capture of new value streams. Unlike before, corporations will be incentivized to curate and structure their scientific data beyond what is required for IP protection and regulatory compliance. Corporations that do not do so will be left at a disadvantage in the market, while competitors soar ahead with better products. *ML will not replace scientists, but scientists who use ML will replace those who do not* [33].

There are several changes we can expect: First, it will be necessary to adopt NoSQL technologies with flexible schemas which can capture the wide range of possible work streams in material science, while at the same time providing the much-needed standardization of materials data. Second, companies will be compelled to annotate data with much more information than "who did what when." Materials data will become discoverable (by relevant scientific criteria), standardized and structured, and ML ready. The later condition will require for domain specific information to be effectively converted to machine-readable data. Chemical structures, microscopy images, spectroscopic data, and X-ray diffraction patterns, to name a few, will require specific transformations to guarantee their interoperability in a machine learning context. Figure 6.14 summarizes these requirements.

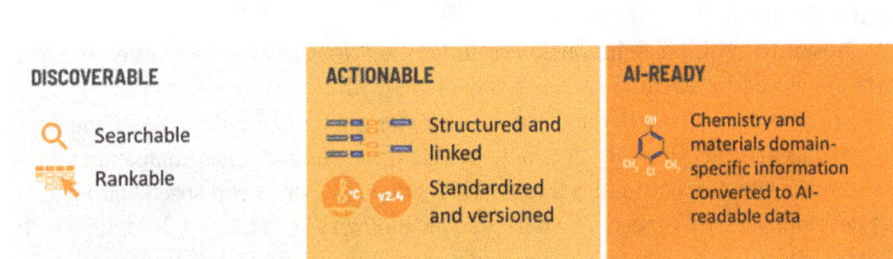

DISCOVERABLE

Searchable

Rankable

ACTIONABLE

Structured and linked

Standardized and versioned

AI-READY

Chemistry and materials domain-specific information converted to AI-readable data

Fig. 6.14: Requirements for ML-ready materials data.

In our experience at Citrine Informatics helping leading materials companies transition to this new paradigm in materials data management, we have found success by using a NoSQL directed acyclic graph-based model, which we developed in-house and have since open-sourced. We dubbed this model GEMD, which stands for graphical expression of materials data [34]. The model aims at capturing all the events (ingredients, processing steps, conditions, etc.) that lead to the generation of a given material. We call such a series of events a material history. GEMD has enabled chemical companies to capture the necessary context in materials data to guarantee its interoperability and reusability while maintaining the required schema flexibility. Discussing the details of the GEMD data model is beyond the scope of this chapter, but as the model is open to everyone, the reader is encouraged to probe further in the references provided [34]. Figure 6.15 shows an example of a GEMD material history.

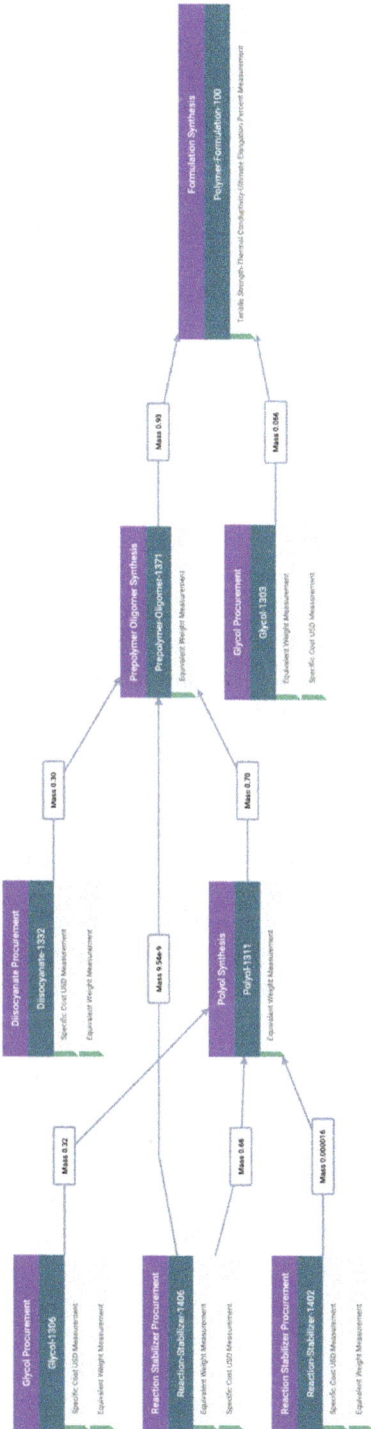

Fig. 6.15: Example material history for the graphical expression of materials data model.

6.4 Into the digital age: challenges and opportunities for the commercial adoption of materials informatics

We described the societal and market conditions compelling chemicals and materials companies to invest in material informatics. We also covered what the current applications of materials informatics are in industrial settings and we took a "short" detour to evaluate the state of the data-management ecosystem. To conclude, we will highlight current challenges and opportunities for wider commercial adoption of MI and present a forward-looking outlook of where the commercial application of materials informatics is likely to go. We will divide the challenges and opportunities intro two areas: culture and data.

6.4.1 Culture

A company's internal culture likely poses the greatest barrier for successfully implementing a new technology like materials informatics. MI is not a simple productivity tool or an IT investment, it is an inherently disruptive process that requires staff to significantly change their daily routines, evolve the way they approach and think about material development problems, and for the corporation to meaningfully evolve their business model. Detractors of data-driven methods and adherents to the traditional R&D process may resist change and become large barriers for wider adoption, specially early on. Ensuring success in a new technological environment will require sufficient training. Skillsets will also have to be re-focused to support data-driven materials development.

We have seen two opportunities to accelerate the cultural change required for the adoption of materials informatics. First, support from senior executives is essential to drive the necessary change management across the organization. Second, early wins in pilot projects and identification of internal "champions" help create momentum to overcome larger cultural barriers.

At a corporate level, chemicals and materials companies might also find it difficult to move away from the asset-intensive business model that has been created, at least partially, by the slow rates of innovation and product differentiation which materials informatics is meant to disrupt. Even as margins erode and competing within the old model becomes more challenging, a World Economic Forum report (2016) highlighted that *46% of global chemical companies have limited means to reposition their capital investments in the next decade, which will lead to a limited willingness to drastically change business models* [8].

Another major area of concern which will affect the adoption of materials informatics within the chemicals and materials industry is its ageing workforce. As of 2016, the median age of workers at US chemical companies was 45.3 years, with 23% of the workforce being eligible for retirement within the next 10 tears [8]. The combination of an ageing workforce and the need for the development of new data-driven skills might be one the biggest challenges for the widespread adoption of materials informatics.

6.4.2 Data

As a field, materials science suffers from an unusually high data-acquisition cost, with a single experiment costing thousands or even hundreds of thousands of dollars. This is of course why experiment prioritization through sequential learning and portfolio rationalization through search space evaluation have found commercial success. However, the lack of sufficient training data remains a major challenge when aiming at scaling materials informatics operations [23, 25]. Most companies may have a small number of curated datasets painstakingly built by industrious scientist through many hours of running parameter sweeps, digging through poorly annotated files and wrangling information into a meaningful data corpus. These datasets are used up first, as companies aim to have quick demonstrations of value. Since data generation is slow and expensive, kickoff of subsequent SL projects in new product lines or business units usually requires a quest in data archeology. Data archeology is distinct from data mining. While data mining focuses on finding correlations in large volumes of structured data, data archeology searches for pockets of valuable data in a large volume of low-quality dirt. The reality for chemicals and materials companies is that not all their historical data is equally useful. Years of neglect have taken a toll on the data corpus and poorly annotated data with little scientific context is particularly difficult to curate into a meaningful training set.

It is not all bad news however, as "the data problem" can be iteratively solved while still generating value. Although kickoff takes some effort on data preparation, successive SL projects generate a growing corpus of high-value, reusable data assets. Valuable data dug from the historical archeological site is progressively added to the ML-ready corpus, increasingly adding value to the data asset. In the meantime, the corporation generates valuable institutional knowledge in materials informatics and is able take the necessary measures to make sure go-forward data curation and storage is done appropriately.

6.5 Outlook: the future of materials informatics

To date, material informatics has been largely focused on materials design and green field discovery at the laboratory scale. However, not every new application requires a radically new material, and there is a huge potential to leverage MI holistically in the rest of the materials lifecycle. Besides discovery of novel materials, we should consider materials informatics for materials selection, optimization for product use, certification, and manufacturing [17].

First, the goal of materials development is, in almost all cases, to enable better product development. As such, we should not neglect the problem of effectively finding the optimal material solution for a product requirement within the space of known materials. The objective of engineers in charge of product development is to select the optimal material, produce that material at scale, and estimate how the material will behave over the product lifetime. Sometimes simply finding the correct existing material for a new application can have major consequences. For example, the super strong and corrosion-resistant allow Iconel (625), initially developed in the 1960s as a structural material for supercritical steam power plants, was identified and introduced in 2016 into the battery contacts of the Tesla® Model S [17]. Iconel (625) happened to be particularly well suited for this application due to its stress response at the very high temperatures that resulted from resistive heating during heavy acceleration.

The task of selecting the best material for a given application can be daunting, especially when the inventory of existing materials is massive.The US Environmental Protection Agency current reported more than 84,000 existing chemicals in its records as of 2015. The Chemical Abstracts Service had over 100 million substances documented in the same year [8].

The task of selecting the correct material for an application is mathematically very similar to the inverse design material discovery problem we have described using sequential learning. Given a large catalog of known materials and a list of performance requirements, we must identify those which are non-dominated, i.e. close to the Pareto front.

In this way, we can envision a paradigm shift in product design. A product engineer could quickly screen for known materials using data-driven methods and interpolate if necessary, co-optimizing the product and material design simultaneously.

Direct integration between product and materials development will soon be possible. Product and part designers will be able to find or quickly formulate new materials tailored for their specific applications. Materials companies will move from offering a catalog of (increasingly commoditized) materials to a portfolio of tailored solutions. Figure 6.16 describes this concept.

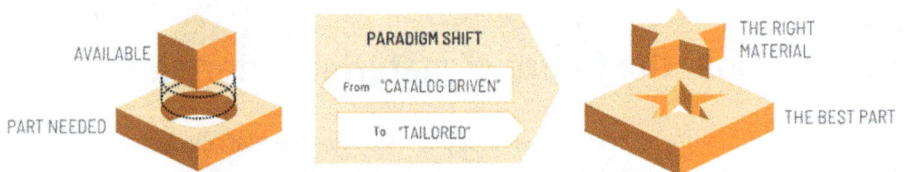

Fig. 6.16: Future paradigm of materials development.

Tab. 6.3: Future state of materials development with MI.

	TODAY	FUTURE STATE WITH MI
MATERIAL AND CHEMICAL SPECIFICATION	Part optimization with new materials is resource-intensive and time-consuming.	Integrated materials and product-level optimization identifies the right material for the application.
RATE OF INNOVATION	New materials development time is a bottleneck in new product development.	Materials-aware AI lets product developers specify the right material, faster.
IP AND CONTROL OVER MATERIAL AND CHEMICAL SELECTION	Product developers have a dearth of materials data and little control over material properties or chemistry.	Product developers identify and own new materials, leading to sustainable competitive differentiation, supply chain flexibility, and optimized cost position.

This paradigm shift will not only debottleneck product development cycles and halt the margin erosion in the materials industry, but it will also enable the industry to tackle some of the largest sustainability problems related to materials development. For example, co-optimization of food packaging and plastic resins will enable us to engineer durable packaging with greatly improved recyclability. Co-optimization of cement and concrete mixtures could lead to improvements in concrete strength and, most importantly, reductions in carbon dioxide generation. Table 6.3 summarizes the future state that materials informatics will enable.

References

[1]   Sparks TD. Discover the materials of the future . . . in 30 seconds or less, TED: Ideas Worth Spreading, 2019.
[2]   De Pablo JJ, Jackson NE, Webb MA, Chen L-Q, Moore JE, Morgan D, Jacobs R, Pollock T, Schlom DG, Toberer ES, Analytis J, Dabo I, DeLongchamp DM, Fiete GA, Grason GM, Hautier G, Mo Y, Rajan K, Reed EJ, Rodriguez E, Stevanovic V, Suntivich J, Thornton K, Zhao J-C. Npj Comput Mater, 2019, 5, 41.
[3]   Paul DR, Robeson LM. Polymer (Guildf), 2008, 49, 3187–3204.
[4]   Gates B. How to Avoid a Climate Disaster, 2021.
[5]   14 Grand Challenges for Engineering in the twenty-first Century, http://www.engineeringchallenges.org/challenges.aspx.
[6]   Hertwich EG. Nat Geosci, 2021, 14, 151–155.
[7]   Allwood JM, Ashby MF, Gutowski TG, Worrell E. Philos Trans R Soc A Math Phys Eng Sci, 2013, 371, 20120496.
[8]   World Economic Forum. Advanced Materials Systems Chemistry and Advanced Materials, 2016.
[9]   Allwood JM, Cullen JM, Milford RL. Environ Sci Technol, 2010, 44, 1888–1894.
[10] Allwood J. Sustainable materials – with both eyes open, 2012.
[11] Czigler T, Reiter S, Schulze P, Somers K. Laying the foundation for zero-carbon cement, 2020.
[12] Carruth MA, Allwood JM, Moynihan MC. Resour Conserv Recycl, 2011, 57, 48–60.
[13] Simons TJ. Chemicals 2025: Will the industry be dancing to a very different tune?, 2017.
[14] Morawietz M, Bäumler M, Caruso P, Gotpagar J. Booz Co, 2010, 17.
[15] Schmitz C. What to do when a specialty-chemical business gets commoditized, 2017.
[16] Simons TJ. Commoditization in chemicals: Time for a marketing and sales response, 2016.
[17] Mulholland GJ, Paradiso SP. APL Mater, 2016, 4, 053207.
[18] Annunziata M. Mind Over Matter: Artificial Intelligence Can Slash The Time Needed To Develop New Materials, https://www.forbes.com/sites/marcoannunziata/2018/12/03/mind-over-matter-artificial-intelligence-and-materials-science/?sh=4115b4dee9db.
[19] Budde F. The state of the chemical industry – it is getting more complex, 2020.
[20] Del Rosario Z, Rupp M, Kim Y, Antono E, Ling J. J Chem Phys, 2020, 153, 024112.
[21] Meredig B, Antono E, Church C, Hutchinson M, Ling J, Paradiso S, Blaiszik B, Foster I, Gibbons B, Hattrick-Simpers J, Mehta A, Ward L. Mol Syst Des Eng, 2018, 3, 819–825.
[22] Ling J, Hutchinson M, Antono E, Paradiso S, Meredig B. Integr Mater Manuf Innov, 2017, 6, 207–217.
[23] Kim Y, Kim E, Antono E, Meredig B, Ling J. Npj Comput Mater, 2020, 6, 131.

[24] Jablonka KM, Jothiappan GM, Wang S, Smit B, Yoo B. Nat Commun, 2021, 12, 1–10.
[25] Peerless JS, Sevgen E, Edkins SD, Koeller J, Kim E, Kim Y, Gargt A, Antono E, Ling J. MRS Commun, 2020, 10, 18–24.
[26] Deloitte. Deloitte Insights.
[27] Hill J, Mulholland G, Persson K, Seshadri R, Wolverton C, Meredig B. MRS Bull, 2016, 41, 399–409.
[28] Silbergeld EK, Mandrioli D, Cranor CF. Annu Rev Public Health, 2015, 36, 175–191.
[29] Waters Corporation. The Necessity of SDMS with LIMS : Reducing Cost and Improving Efficiency in the Analytical, 2011.
[30] www.atriumresearch.com, Smart Lab : Laboratory Informatics Exchange, 2006.
[31] Alberi K, Nardelli MB, Zakutayev A, Mitas L, Curtarolo S, Jain A, Fornari M, Marzari N, Takeuchi I, Green ML, Kanatzidis M, Toney MF, Butenko S, Meredig B, Lany S, Kattner U, Davydov A, Toberer ES, Stevanovic V, Walsh A, Park N-G, Aspuru-Guzik A, Tabor DP, Nelson J, Murphy J, Setlur A, Gregoire J, Li H, Xiao R, Ludwig A, Martin LW, Rappe AM, Wei S-H, Perkins J. J Phys D Appl Phys, 2019, 52, 013001.
[32] Big Data in Materials Research and Development. National Academies Press, Washington, D. C., 2014.
[33] Meredig B. Chem Mater, 2019, 31, 9579–9581.
[34] Citrine Informatics. GEMD Documentation, https://citrineinformatics.github.io/gemd-docs/.
[35] Hutchinson ML, Antono E, Gibbons BM, Paradiso S, Ling J, Meredig B. arXiv.

About the editor

Phil De Luna is currently the youngest-ever Director at the National Research Council of Canada where he leads a world-class $57M research program on Canada-made clean energy technology. His mandate is to develop transformative technologies to help Canada achieve net-zero GHG emissions by 2050.

De Luna is a carbontech innovator with experience spanning cutting-edge topics such as carbon capture and conversion technologies, clean hydrogen, and artificial intelligence for materials discovery. His research has been published in high-impact journals such as *Nature* and *Science* (>6,000 citations in 5 years) and has been covered by mainstream media such as Newsweek, CBC, Forbes, Popular Science, and more. He holds a PhD from the University of Toronto in Materials Science and Engineering, where he was a Governor General Gold Medalist.

De Luna was a candidate for the Green Party in Toronto-St. Paul's in the 2021 Federal Election where he ran to bring more diversity to parliament and more science to politics. He finished in the top 5 percentile of Green Party candidates with a vote share that was 2.7 times greater than the national average and raised the most money ever in his riding.

De Luna was named to the 2019 Forbes Top 30 Under 30 – Energy list and was a finalist (1 of 10 worldwide) in the $20M Carbon XPRIZE. He is Vice Chair of the board of directors of CMC Research Institutes, a carbontech non-profit, a member of the OECD Advanced Materials steering committee, and a member of the Working Group on AI Ethics and Sustainable Development Goals at the Canadian Commission for UNESCO. He is also a Mission Innovation Champion for Canada, an Action Canada Fellow, a Creative Destruction Lab Mentor, and a 2x TEDx Speaker. In his spare time, Phil hosts and produces the podcast "What's Next In . . ." about the rapidly changing world and how we can get ahead of it.

https://doi.org/10.1515/9783110738087-007

Index

https://doi.org/10.1515/9783110738087-008

www.ingramcontent.com/pod-product-compliance
Lightning Source LLC
Chambersburg PA
CBHW061418210326
41598CB00035B/6261